Superior 39.95

WIRELESS
CRASH COURSE

McGRAW-HILL
TELECOMMUNICATIONS

Wireless
Crash Course

Paul Bedell

McGraw-Hill
New York · Chicago · San Francisco · Lisbon
London · Madrid · Mexico City · Milan · New Delhi
San Juan · Seoul · Singapore
Sydney · Toronto

Library of Congress Cataloging-in-Publication Data

Bedell, Paul.
Wireless crash course / Paul Bedell.
p.   cm.
Includes bibliographical references and index.
ISBN 0-07-137210-5
1. Wireless communication systems.   I. Title.
TK5103.2.B43   2001
384.5—dc21   00-054885

McGraw-Hill

A Division of The McGraw·Hill Companies

3 4 5 6 7 8 9 0   DOC/DOC   0 7 6 5 4 3 2 1

ISBN 0-07-137210-5

The sponsoring editor for this book was Stephen S. Chapman, the editing supervisor was Stephen M. Smith, and the production supervisor was Sherri Souffrance. It was set in Vendome ICG by Paul Scozzari of McGraw-Hill's Hightstown, N.J., Professional Book Group composition unit.

Printed and bound by R. R. Donnelley & Sons Company.

Portions of this book were previously published in *Cellular/PCS Management,* © 1999.

McGraw-Hill books are available at special quantity discounts to use as premiums and sales promotions, or for use in corporate training programs. For more information, please write to the Director of Special Sales, Professional Publishing, McGraw-Hill, Two Penn Plaza, New York, NY 10121-2298. Or contact your local bookstore.

This book is printed on recycled, acid-free paper containing a minimum of 50% recycled, de-inked fiber.

In memory of my parents, Robert and Lou Bedell.

To my wife Paula, for her support and understanding the second time around.

And to my special gift from God, my sons Aaron, Ryan, and Robert.

CONTENTS

Contents

Contents

Contents

Contents

Contents

Contents

PREFACE

This book is intended for persons who have a general understanding of the fundamentals of telephony and basic communication systems. A person enrolled in an undergraduate or graduate program in telecommunications should be able to grasp all of the concepts contained within this book.

A layperson who is just starting out in the wireless industry could likewise read this book to gain an excellent understanding of all the operational elements and design fundamentals that comprise a carrier-class wireless communications system.

A prerequisite for reading this book should be one of three items:

1. The reader should have taken an introductory course on voice (or data) communication networks.

2. The reader should have read an introductory text on voice (or data) communications.

3. The reader should have accumulated 2 to 4 years experience in some facet of the telecommunications, data communications, or computer networking industry.

This text could be used by, but is not intended for, engineering students. The approach taken in this book is to deliberately avoid formulaic writings and theories. Should the reader wish to explore the more technical aspects of systems design, it is suggested that one of the William Lee texts on wireless systems design be selected as ideal reading material.

—PAUL BEDELL

ACKNOWLEDGMENTS

I would like to acknowledge the following wireless and telecom industry professionals for their assistance in supplying and verifying information contained in this book, or for reviewing certain chapters to ensure accuracy and completeness:

- Thanks to Ric Biederwolf of Crown Castle, a person who spent years engineering cellular and PCS networks, for always being willing to answer my queries on any number of wireless topics.

- Thanks to Julius Jackson of U.S. Cellular Corporation, who's spent over 25 years in the wireless industry as a microwave radio expert, for his guidance and expertise regarding microwave radio systems.

- Thanks to Dick Jones of U.S. Cellular Corporation, a veteran of the design and launch of the very first cellular system in the United States (Ameritech Mobile), for his initial and ongoing guidance and mentoring.

- Thanks to Rhonda Wickham, Editor-in-Chief of *Wireless Review* magazine, a savvy and knowledgeable industry player, for her guidance regarding and review of the wireless data chapter.

- Thanks to Bob Keeger of Western Wireless, another veteran of the networking and launch of Ameritech Mobile, for his update on current topics relating to interconnection to the PSTN.

- Thanks to Jon Sadler of Tellabs, a former partner and data guru, for his contribution of the new, revised, rewritten wireless data chapter.

- Thanks to all my students in TDC 512 at DePaul University in Chicago, who keep me on my toes and ensure that I maintain my knowledge of the evolution of the exciting wireless industry.

- Finally, thanks to Drs. Greg Brewster and Helmut Epp of DePaul University in Chicago, for their initial and ongoing support during the development of the cellular and wireless course in the Telecommunications Graduate Program, which led to the development of this book.

INTRODUCTION

This book is intended to provide the reader with a comprehensive understanding of the design, operations, and management of wireless communication systems. To that end, the perspective given is that of a wireless communications company, whether it be a cellular carrier or a PCS (Personal Communication Services) carrier. The goal is to present the material in a survey format, an overview. All of the core and ancillary systems that comprise a carrier-class wireless network are topics that are covered in this book. The motivation behind this book is to deliver the information at a fairly high level, yet provide enough detail and technical information so that readers who have a basic grasp of telecommunication systems fundamentals can feel as though they have sufficiently expanded their knowledge base. To underscore this point, the reader should remember that there have been entire books written on many of the topics covered in this book, such as ANSI-41, PCS, microwave radio systems, wireless data, and radio-frequency (RF) engineering. The focus of this book is broad and general, yet it touches on every single aspect of wireless systems design. From the first chapter to the last chapter this book covers many topics that have not been covered to date in most cellular and wireless books.

All elements of wireless systems are explored in a condensed manner, to allow the reader to obtain a base knowledge of all system components, ranging from RF propagation, to towers, to PSTN interconnection, to commercial and business issues.

An electrical engineering degree is not a prerequisite for reading this book. Formulas reflecting RF propagation theory and design have intentionally been left out. Formulas reflecting electrical characteristics of antennas have also been intentionally omitted. This book is written in plain English. The chapters are laid out in a very logical, steppingstone manner: The reader is walked through systems design from the cell site acquisition phase to the commercial and business issues phase. If the reader is seeking to understand how wireless communication systems are designed, built, and approached from the ground up, this is the book to read.

WIRELESS
CRASH COURSE

History of Radio Communications

A gentleman named Heinrich Hertz was the discoverer of electromagnetic waves, the technical foundation of radio itself. By 1880, Hertz had demonstrated a practical radio communication system. This is the origin of the term *hertz*, the unit of frequency.

Guglielmo Marconi developed the world's first commercial radio service in 1898. His first customer was Lloyd's of London, and the first radio link covered about 7.5 mi and provided information about incoming shipping. This link was a *data* communications–only link. It was the first ship-to-shore communications system.

The first human voice transmission via radio was accomplished by Reginald Fessenden in December 1900. This first *voice* radio link was 1 mi long. The demonstration took place in Maryland, and marked the beginning of radio telephony. The first cellular telephone system didn't go into operation until 83 years later.

In 1901, Marconi produced the first long-distance transatlantic radio transmission.

On Christmas Eve in 1901, Fessenden transmitted the world's first radio broadcast. The transmitter was located at Brant Rock, Massachusetts, and good-quality voice and music was received by ship and shore operators within 15 mi of Brant Rock.

From 1910 to 1912 *mandatory* 24-h ship-to-shore communications were established by the United States, Great Britain, and other maritime nations as a direct result of two ships sinking: the *Republic* in 1909 and the *Titanic* in 1912. This requirement was derived from the first attempt at regulation of the radio industry: the Radio Act of 1910.

Key: The Radio Act of 1910 was the first instance of government regulation of radio technology and services. The original act was approved in June 1910 and required certain ocean-going ships, of all nationalities, to carry radio equipment when visiting U.S. ports and to exchange messages with other vessels, regardless of the system used. The original act applied only to ocean-going vessels, and also only required a single radio operator. In July 1912, the original act was amended. Among the changes were the inclusion of vessels on the Great Lakes, coverage of all ships licensed for 50 passengers and crew, and a requirement for a continuous watch, with at least two operators. Neither act required station licenses.

In 1915, a team of Bell Telephone engineers, using the giant antennas at the U.S. Navy station at Arlington, Virginia, were the first to span an ocean with the human voice. This was a milestone in international radio telephony as voice radio transmissions were received in France, Panama, and Hawaii. By 1918, 5700 ships worldwide had wireless telegraphy installations.

1.1   Mobile Radio Systems

The development of the mobile radio system can be divided into two parts: Phase I produced the earliest systems, and Phase II began after the Federal Communications Commission's (FCC) classification of what it termed "Domestic Public Land Mobile Radio Service."

The need to increase public safety was key to the genesis of today's rapidly growing wireless communications industry. The first use of mobile radio in an automobile instead of a ship was in 1921. The Detroit Police Department implemented a police dispatch system using a frequency band near 2 MHz. This service proved so successful that the allocated channels in the band were soon utilized to the limit. In 1932, the New York Police Department also implemented the use of the 2-MHz band for mobile communication.

But the technology to enable mobile communication services for public safety agencies was not yet available. Early radiotelephone systems could be housed on ships with reasonable ease, but were too large and unwieldy for cars. Also, bumpy streets, tall buildings, and uneven landscapes prevented successful transmission of the radiotelephone signals on land. The key technological breakthrough came in 1935, when Edwin Armstrong unveiled his invention, frequency modulation (FM), to improve radio broadcasting. This technology reduced the required bulk of radio equipment and improved transmission quality.

In 1934, the FCC allocated four new channels in the 30- to 40-MHz band, and by the early 1940s a significant number of police and public service radio systems had been developed. By the late 1940s, the FCC made mobile radio available to the private sector, along with police and fire departments.

1.1.1   Mobile Telephone Service (MTS)

In 1946, Bell Telephone Labs inaugurated the first mobile system *for the public*, in St. Louis. This system was known as Mobile Telephone Service (MTS). Keep in mind that at this time AT&T still owned and operated the majority of the public switched telephone network (PSTN). Three channels in the range of 150 MHz were put into service, operating at frequencies between 35 and 44 MHz. An MTS highway system to serve the corridor between Boston and New York began operating in 1947. MTS transmissions (from radio towers) were designed to cover a very large area, using high-power radio transmitters. Often the towers were placed at geographically high locations. Because they served a large area, they were subject to noise, interference, and signal blocking.

MTS was a half-duplex, "push-to-talk" system; therefore MTS offered communications that were only one way at a time. An operator was needed to connect a customer to the landline local exchange carrier (LEC) network.

In 1949 the FCC authorized non-wireline companies known as radio common carriers (RCCs) to provide MTS. An RCC is a wireless carrier that is not affiliated with a local telephone company. Prior to 1949, all mobile service was supplied by the *wireline* telephone companies. This marked the birth of competition in the telecommunications industry.

1.1.2 Improved Mobile Telephone Service (IMTS)

In 1965, almost 20 years after the introduction of MTS, the Bell System introduced Improved Mobile Telephone Service (IMTS), the successor system to MTS. IMTS was the first automatic mobile system: it was a full-duplex system, eliminating the push-to-talk requirement of the older MTS system. IMTS allowed simultaneous two-way conversations. A key IMTS advantage was that users could dial directly into the PSTN. IMTS narrowed the channel bandwidth, which increased the number of frequencies allowed. Because the cell site locations were high-output-power stations, one radio location could serve an entire city.

Between the landline phone company and the RCC, nineteen 30-kHz channels were authorized in the 30- to 300-MHz band, which is the VHF band. The FCC also authorized twenty-six 25-kHz channels in the 450-MHz band (the UHF band). With full-duplex systems such as IMTS, two radio channels are needed for each conversation: one channel to transmit and one channel to receive.

As with MTS, IMTS radio towers were still installed in high places (e.g., tall buildings), and the system was still designed to cover large geographic areas, up to 50 mi in diameter. Because of limited capacity, eventually IMTS operators prohibited roaming in their markets. Roaming refers to placing calls in markets other than a user's home market. Roaming will be discussed in a later section.

Trivia: The IMTS system was designed so that only 50% of the calls were completed during the busy hour. Service was often poorer than that in some metropolitan areas. This was a result of the fact that very few radio channels existed for IMTS service.

1.2 AMPS: The American Cellular Standard

In an effort to use the airwaves more efficiently, AT&T engineers decided to stretch the limited number of radio frequencies available for mobile service by scattering multiple low-power transmitters throughout a metropolitan area, and "handing off" calls from transmitter to transmitter as customers moved around in their vehicles. This new technique would allow more customers access to the systems simultaneously, and when more capacity was needed, the area served by each transmitter could be divided again. This was the birth of wireless technology as we know it today.

Advanced Mobile Phone Service (AMPS) is the American analog cellular standard. In 1970, several key developments occurred:

1. The FCC set aside new radio frequencies for land-mobile communications. These frequencies were UHF television channels in the 800-MHz band that had never been used.

2. That same year, AT&T proposed to build the first high-capacity cellular telephone system. It dubbed the system AMPS for Advanced Mobile Phone Service and selected Chicago as the first test city.

At the inception of the cellular industry, the FCC initially granted a total of 666 channels in each market. At first, AT&T thought it would get national rights to *all* cellular frequencies, thereby making AT&T the only national cellular carrier. This would also have made the cellular industry a monopoly.

Trivia: At that time, AT&T never anticipated the growth potential of the apparently pent-up demand by the general public for widespread availability of mobile communication services. They estimated only 1 million cellular customers would exist by the end of the century. Today, there are over 100 million wireless customers in the United States alone!

However, at that time the FCC, bowing to intense pressure from radio common carriers (RCCs), determined that the cellular industry should have two carriers per market, and 333 channels were allocated per carrier, per market. This marked the birth of the A band and B band carrier concept (see Chapter 2, "The Cellular Market Regulatory Structure"). The number of channels was later increased to a total of 832 total cellular channels, *416 channels per carrier* per market. This change was brought about by cellular industry pressure on the FCC to relinquish reserve spectrum to relieve capacity and congestion problems.

In 1977, while the FCC realized it had to create a regulatory scheme for the new service, the Commission also decided to authorize construction of two developmental cellular systems: one in Chicago licensed to Illinois Bell, and a second serving Baltimore and Washington, D.C., licensed to a non-wireline company: American Radio Telephone Service (an RCC).

Once the regulatory framework was decided upon by the FCC, the first commercial cellular system began operating in Chicago on October 13, 1983. The very first commercial cellular telephone call was made at Soldier Field in Chicago to a descendant of Alexander Graham Bell in West Germany. The second system was activated a short time later in the Baltimore/Washington, D.C., corridor in December 1983. It was these systems that gave rise to the fastest-growing consumer technology in history, an industry that adds about 28,000 customers per day.

1.3 Definition of Cellular Radio

There are two different ways to view the definition of cellular systems:

FCC definition A high-capacity land mobile system in which assigned radio spectrum is divided into discrete channels which are assigned in groups to geographic cells covering a cellular geographic service area (CGSA). The discrete channels are capable of being reused in different cells within the service area, through a process known as *frequency reuse.*

Layman's definition   A system which uses radio transmission rather than physical wirelines to provide telephone service comparable to that of regular business or residential telephone service.

1.3.1   The Cellular Concept

Instead of having just a few radio channels that everyone must share [like MTS, IMTS, or citizens' band (CB) radio], cellular radio channels are reused simultaneously in nearby geographic areas, yet customers do not interfere with each other's calls.

The cellular system is similar in functional design to the public switched telephone network, or landline network: fundamentally it contains subscribers, transmission systems, and switches. The existence and control of the radio function of the cellular system is what differentiates cellular from landline telephone service (the PSTN). When launched in 1983, the cellular radiotelephone system was the culmination of all prior mobile communication systems.

1.4   Cellular System Objectives

When Bell Labs developed the AMPS cellular concept, the major system objectives were efficient use of radio spectrum and widespread availability. While the cellular concept had been developed earlier, several critical technologies came together simultaneously in the late 1970s to propel the cellular industry forward. New technologies enabled small, relatively lightweight subscriber equipment to be manufactured cheaply. Vastly improved integrated circuit manufacturing techniques allowed for major advances in computer technology as well as the miniaturization of critical equipment elements, especially within portable mobile telephones.

Key: There are *five main components* to the cellular telephone system: the mobile telephone, the cell base station, the mobile switching center (MSC), the fixed network (transmission systems), and the PSTN. See Figure 1-1.

Figure 1-1
Five main
components of
cellular/wireless
communication
systems.

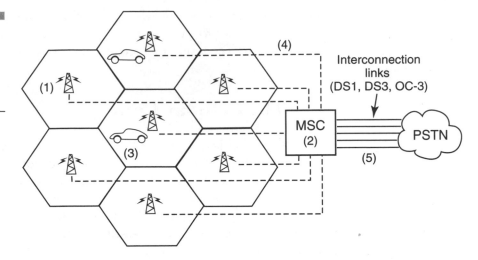

(1) ⚡🗼⚡ = Base transceiver station (BTS)

(2) MSC = Mobile switching center

(3) ⌐⊙⊙⊃ = Mobile unit (mobile station)

(4) - - - - - - = Fixed network

(5) (─⊂⟩) = Interconnection to the PSTN

1.5 Test Questions

True or False?

1. _____ Mobile Telephone Service (MTS) transmissions were designed to cover a small area, using low-power radio transmitters.

2. _____ Mobile Telephone Service (MTS) was the first automatic mobile system, featuring full-duplex functionality allowing simultaneous two-way conversations.

3. _____ In 1977, the FCC authorized construction of two developmental cellular systems: one in the Baltimore/Washington, D.C., corridor and one in Seattle.

Multiple Choice

4. Who discovered electromagnetic waves, which are the foundation of radio systems?
   (a) Guglielmo Marconi
   (b) Reginald Fessenden
   (c) Harold Hertz
   (d) Heinrich Hertz
   (e) The Bell Telephone System

5. The first use of mobile radio in an auto versus a ship was:
   (a) By the New York Police Department in 1932
   (b) By the U.S. Navy in 1915
   (c) By Reginald Fessenden in 1906
   (d) By the Detroit Police Department in 1921

6. Initially, how many channels, in what frequency range, were allocated to Mobile Telephone Service (MTS)?
   (a) 19 30-kHz channels in the 30- to 300-MHz band
   (b) 17 channels in the 1.9-MHz range
   (c) 75 channels in the 800-MHz range
   (d) 3 channels in the 150-MHz range

2

The Cellular Market Regulatory Structure

2.1 MSAs and RSAs

The Federal Communications Commission (FCC) divided the United States into 734 cellular markets. The larger, metropolitan markets are known as metropolitan statistical areas (MSAs). The original intent of the FCC was that they be geographic areas containing a population of 150,000 or more persons. There are 306 MSAs in the United States, and they are labeled according to the largest city within their boundaries.

The smaller, rural markets are known as rural service areas (RSAs). There are 428 RSAs in the United States. The original intent of the FCC was that they be geographic areas containing a population of less than 150,000 persons. RSAs are labeled according to their state, beginning with the number 1, from the north or west side of the state, progressing in sequence to the south or east end of the state. For example, DeKalb, Illinois, is located in Illinois RSA 1 at the north end of the state, and Kentucky RSA 1 is located in the western end of the state of Kentucky.

The FCC developed and demarcated MSAs first, and then a few years later it developed and demarcated RSAs. MSAs and RSAs run along county lines, as they are defined by Rand McNally. Although some markets (usually MSAs) may be defined around one county, most cellular markets encompass more than one county. See Figure 2-1.

Figure 2-1
MSA and RSA
market layout.
(Geographic boundaries are not to scale.)

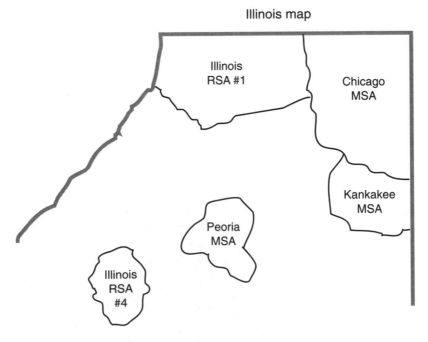

Illinois map

Illinois
RSA #1

Chicago
MSA

Kankakee
MSA

Peoria
MSA

Illinois
RSA
#4

Key: MSAs cover 75% of the population and 20% of the landmass of the United States, while RSAs cover 25% of the population and 80% of the landmass.

In 1983, the FCC decreed that cellular carriers had to provide coverage to *either* 75% of the population of a given market or 75% of the geographic area of a given market. This rule gave carriers the freedom to build out their markets more efficiently, as they saw fit. For example, they could concentrate their system buildouts in MSAs to just the urban area itself, in order to serve as many customers as possible as quickly as possible. Conversely, in RSAs, they could spread the buildout throughout the market's area to serve as many people as possible in as short a time frame as possible.

2.2   The A Carrier and B Carrier Designations

In 1981, the FCC released its *Final Report and Order* on cellular systems, specifying there would be two competing cellular companies licensed in each market. Because of this FCC specification, the cellular industry is a duopoly by decree. Once its rules were in place, the FCC accepted applications for the 60 largest cities during 1982. The Commission decided to license cellular systems in the 306 MSAs first, and then move on to the 428 RSAs. The FCC decreed that one license in each market would be reserved for the local telephone company. This is the so-called wireline license, and the licensee is known as the B carrier in that market. The B stands for Bell, indicating the (B) carrier's association with the local telephone company. The second licensee in cellular markets is known as the A carrier. The A stands for alternative, and marks the (A) carrier as the non-wireline carrier, indicating that this carrier has no association with the local telephone company in that particular market. The FCC set up the cellular regulatory framework in this way for a reason: It ensured that at least one licensee in each market would have telecommunications industry experience, and would already have some type of telecommunications infrastructure in place.

Key: The distinction between landline telephone companies as the B carrier in cellular markets and non-wireline companies as the A carrier in cellular markets has become blurred and unimportant over the years. Today, any qualified company can obtain either the A or B license in any MSA/RSA.

2.3   Initial Cellular Licensing

The FCC decided initially to conduct comparative hearings to select the most qualified applicants for the A and B block licenses when more than one company applied for the same license. "Comparative" means that the FCC examined the prospective licensee's proposed coverage and growth plans. In early 1983, 567 applications were filed just for the 30 markets ranked 61st to 90th in size! The FCC quickly realized that comparative hearings were too slow and a new licensing process needed to be found. It was then taking 10 to 18 *months* of deliberation and more than $1 million in costs to award a single license. So in October 1983, the FCC amended its rules and specified that lotteries would be used to select the A and B carriers in cellular markets among competing applicants in all but the top 30 markets. The lottery winners faced a detailed legal, technical, and financial review before the FCC would issue a license to construct a cellular system. A key purpose in conducting this review was to ensure that the prospective licensees had a documented build plan that made sense. By 1990, construction permits had been issued for at least one system in every cellular market in the United States.

It is important to note that the terms cellular market, cellular system, and MSA/RSA are all synonymous and interchangeable.

Key: When a cellular carrier has multiple cellular markets that abut each other in a geographically contiguous area (sharing common borders), this is known as a *cellular cluster.*

2.4   Test Questions

True or False?

1. _____ MSAs are cellular markets containing populations of 150,000 or less persons.

2. _____ Cellular market (RSA/MSA) boundaries run along county lines, as they are defined by Rand McNally.

3. _____ The distinction between landline telephone companies as the B carrier in cellular markets and non-wireline companies (radio common carriers) as the A carrier in cellular markets has become blurred and unimportant in recent years.

4. _____ The "A" in "A Carrier" stands for the word *access*.

5. _____ Cellular RSAs are labeled according to the largest city within their boundaries.

Multiple Choice

6. The FCC created the cellular industry as a "duopoly," with an A carrier and a B carrier existing in each cellular market. At the inception of the cellular industry, the FCC decreed that the B carrier would be associated with the local telephone company because:
   (a) AT&T wanted it that way.
   (b) The A carriers insisted that a local telco should be their chief competition.
   (c) This structure ensured that at least one cellular licensee in each market would have telecommunications industry experience and would already have some type of telecommunications infrastructure in place.
   (d) A similar market structure was already in place in Japan, and the FCC was following that country's lead.
   (e) Motorola recommended this structure to the FCC.

Fundamental Wireless System Design and Components

3.1 Frequency Reuse and Planning

The wireless system enables mobile communications through the use of a very sophisticated two-way radio link that is maintained between the user's wireless telephone, the wireless network, and the landline public telephone network. The concept behind the two-way radio link is ingenious. It involves using individual radio frequencies over and over again throughout a city or county with minimal interference, to serve a large number of simultaneous conversations. This concept is the central tenet of cellular system design and is known as *frequency reuse*. The frequency-reuse concept is what separates the cellular system from all preceding systems such as MTS and IMTS.

The major drawback with previous mobile communication systems was the inefficient use of allocated radio spectrum. *Reusing* radio frequencies over a given geographic area provides the means for supporting a number of concurrent conversations far in excess of the number of voice channels derived from simply parceling out available spectrum, as was done with MTS and IMTS. The proximity of existing cell base stations and the existing frequency-reuse plan *must* be taken into account when examining where to build new cells into an existing system. The frequency-reuse plan is defined as how radio-frequency (RF) engineers (working for wireless carriers) subdivide and assign the FCC-allocated radio spectrum, throughout the carrier's market.

Key: The main challenge of cellular system engineering is to grow a cellular system in capacity by adding cell base stations and radio channels (*to existing cells*) so that the base stations throughout a system do not interfere with each other.

A frequency-reuse plan is produced in groups of 7 cells. $N = 7$, where N is the number of cells in the frequency-reuse plan. There are *many* 7-cell frequency-reuse groups in each cellular carrier's MSA or RSA. What this means is that all available 416 channels (conversation frequencies) in a cellular market are plotted throughout the 7-cell frequency-reuse plan over and over again. Higher-traffic cells will have more radio channels assigned to them in order to optimize the capacity of the system according to customer usage. See Figure 3-1.

Figure 3-1
Frequency reuse: N =
7 reuse format.

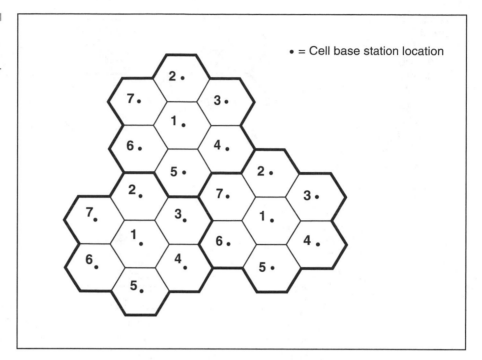

• = Cell base station location

3.2 Distance-to-Reuse Ratio

The distance-to-reuse ratio (D/R) defines the geographic distance that is required between cells using *identical frequencies,* in order to avoid interference between the radio transmissions at these cells. The D/R and ratio is actually a complicated mathematical formula that is derived from extensive field research and testing. The distance-to-reuse ratio is a critical cellular radio principle, and a core element of cellular system design. The overall geographic coverage size of cell base stations *throughout a system* determines the actual distance-to-reuse ratio, along with the radio power level transmitted from cell base stations transmitting the same set of $N = 7$ frequencies. See Figure 3-2.

The frequency agility concept goes hand in hand with the D/R and frequency-reuse concepts. Frequency agility is the capability of the mobile phone to operate on any given frequency in the cellular radio spectrum. The mobile telephone has the ability to change channels almost instantly upon command from the wireless system—from the MSC. Frequency agility can also be described as the ability of the mobile telephone to perform a seamless transfer operation from one radio channel to another.

Figure 3-2

The size of the coverage area and radio power level transmitted from cell base stations transmitting the same set of (N = 7) frequencies dictates the distance-to-reuse ratio in a wireless system.

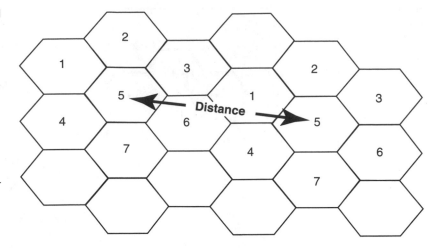

Frequency agility allows for very large use of the relatively small amount of radio spectrum that's been assigned to the cellular industry by the FCC. This large increase in the capability of the allotted amount of spectrum is known as *trunking gain*. This capability also allows for another key cellular system concept, known as *call handoff.*

3.3   Call Handoff

What makes the cellular system work at all is its ability to make the subscriber unit (the mobile phone) change frequency (channel) as the unit moves throughout a market. This is known as *call handoff.* Call handoff is the process where a call *in progress* is transferred from one cell base station to another cell (via a channel change) while maintaining the call's connection to the cellular system (the switch—MSC). Along with frequency reuse, call handoff capability is the driving force behind wireless technology. The call handoff process goes hand in hand with the concept of frequency reuse. Without call handoff, frequency reuse would not be technically feasible.

The rationale behind the handoff process is the need to keep the subscriber unit receiving a usable signal. As the MSC sees a mobile's signal level going down, it looks for a neighboring cell base station that can "hear" the subscriber better via stronger received signal. The call is then handed off to that neighbor cell base station.

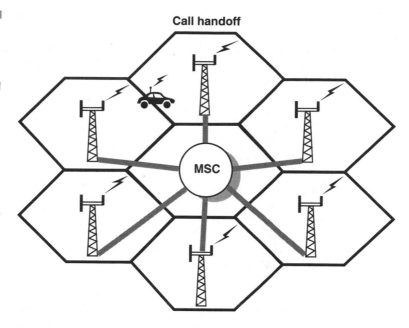

Figure 3-3
In-progress cellular calls are transferred from one cell base station to another cell base station while maintaining the call's connection to the cellular system. All call handoffs are coordinated by the mobile switching center (MSC).

Key: Base station radio coverage from each cell *overlaps* with all the cells that are adjacent to that cell. This overlapping radio coverage is what actually allows for call handoff to occur.

The call handoff process is microprocessor-controlled, like most wireless call processing. Ideally, the call handoff process is *transparent* to the user. The call handoff feature accounts for much of the complexity within a cellular radio handset (i.e., the frequency synthesizer and the control and memory functions).

Call handoff demonstrates why all mobile telephones must have *frequency agility,* the ability to change from one channel (frequency) to another. Call handoff is known as *call handover* in other parts of the world, especially Europe. See Figure 3-3.

3.4   The Hexagon Grid

To facilitate the design of the cellular system with frequency reuse in mind, cells are laid out as hexagons for frequency reuse design purposes.

Figure 3-4
Hexagon grids are
ideal cellular system
design and engineer-
ing tools because
they graphically and
functionally depict
overlapping radio
coverage between,
and among, adjacent
cell base stations.

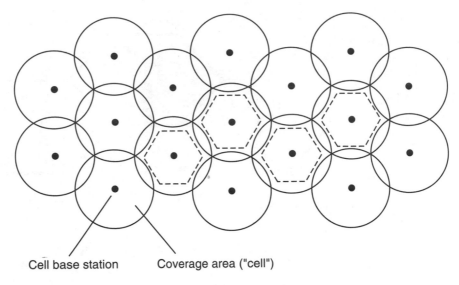

Cell base station   Coverage area ("cell")

This is known as the "hex metaphor." The hexagon, or hexagon grid, is the predominant engineering design tool in the wireless industry. The "hex" grid is printed on a large, clear plastic sheet, and allows engineers to visualize a wireless system for design purposes, regardless of the actual terrain in a given market. Hex grids are laid over United States Geological Survey maps when engineers design wireless systems. The hexagon is used because it best simulates a *grid of overlapping circles*. The overlapping circles represent neighboring cell base stations in a wireless system. (See Figure 3-4.) The hex grid can contain hexes of many different sizes, depending on the density of a given wireless market (system). Urban markets such as Chicago or L.A. will have smaller hex grids since there will be more base stations in these urban markets because they have a more dense population base, and therefore more (installed) cell base stations overall. Base stations in rural areas cover a larger geographic area than base stations in urban areas because the populations in rural areas are slimmer and more dispersed. (See Chapter 4, "The Cell Base Station.") Call handoff *cannot* occur unless there is some overlapping of radio coverage between any given cell site and all its neighboring cell sites. The wireless switch (the MSC), tracks the radio coverage (power levels) of all wireless phone calls. When a call handoff is required, the switch knows which surrounding sites are the best candidates to receive the call handoff based on power measurements taken by the MSC. The key benefit of using the hex grid is that it allows a cellular carrier to grow in an orderly fashion in terms of coverage and the frequency-reuse plan.

For design purposes, hexes can be laid out on paper with the points facing up or any flat side facing up. This depends on the preference of cellular engineers or the actual configuration of the cellular system being addressed. Hex layouts can all be modified according to the needs of the particular cellular system. The hex grid allows a wireless carrier to design a system to reduce or eliminate signal interference and facilitates long-term frequency planning. This includes cells within your own market, as well as cells at the border of your market and another carrier's market. Failure to use a grid system for planned growth makes it harder to add additional cells into a wireless market without *retuning* the radios at existing cells. The hex grid also facilitates cellular design in terms of identifying call handoff candidate cells for every given cell in the system. This information is stored in the MSC. The hex grid can also be used for *traffic analysis* purposes.

3.5 Fundamental Cellular System Components

3.5.1 The Mobile Telephone

There are several types of wireless telephones in use today:

1. *Mobile telephones.* These phones are stationary, nonremovable phones, mounted in automobiles.

2. *Transportable telephones.* These types of cellular telephones are virtually extinct today. They represented the initial migration to truly "portable" phones in that they could be carried with the user anywhere. This type of phone was known as a *bag phone,* and was carried in a handbag-sized carrying case, or used in the car with a special cigarette lighter adapter that enabled the user to conserve the internal battery of the phone. These phones were typically big, bulky, and inconvenient. They were sometimes called "brick" phones due to their weight, appearance, and bulk. They were popular circa 1992—1993. They were quickly displaced in the cellular marketplace by portable phones, because of the portable's light weight and small size.

3. *Portable telephones.* Users are able to carry these cellular telephones with them anywhere. They have rapidly become the most popular form of cellular telephone. They are small and relatively lightweight, and sometimes referred to as *pocket telephones*. Since the mid-1990s, manufacturers have made great breakthroughs in producing portable wireless phones that are very small and very lightweight. A marketing battle has been waged by mobile telephone manufacturers to see who can make the smallest, lightest portable phone. In the late 1990s, Motorola made a marketing breakthrough with its "Star Tac" phone, which was about the size of a deck of playing cards. Another breakthrough occurred in the late 1990s when the Swatch company came out with a wrist phone, although this type of phone didn't really catch on in the marketplace.

4. *Digital cellular phones.* As wireless carriers migrated to digital radio systems, digital wireless telephones are now ubiquitous in the marketplace. These phones were smaller and more streamlined than analog portable telephones, when introduced in the late 1990s.

5. *Dual-mode phones.* Dual-mode wireless phones are those phones that can operate in either analog or digital modes. Today, all digital cellular phones are also dual-mode phones. If a customer were to purchase a *digital* phone, but felt that this phone didn't perform very well in terms of its ability to process a quality digital call, the user could manually program the phone to operate in *analog* mode. As wireless carriers continue the migration to digital radio technology, dual-mode wireless phones have become commonplace. With the advent of Personal Communication Services (PCS), the term *dual mode* will be taking on another meaning. Since PCS carriers are selecting and deploying different *digital* radio technologies, dual-mode phones that will operate in two different digital modes are being developed (e.g., GSM and CDMA), along with phones that will operate in analog and digital mode in the same frequency band (e.g., 800 MHz for AMPS). Digital radio technologies will be discussed in a later chapter.

3.5.2   The Cell Base Station

The cell base station is the location of the radios and tower that serve cellular subscribers (customers): persons who are making and receiving mobile telephone calls.

3.5.3   The Fixed Network

The fixed network is the web of transmission systems that connect all cell base stations to the MSC, the wireless switch. This network is what effectively connects the mobile subscriber—the customer—to the outside world, and vice versa.

3.5.4   The Mobile Switching Center (MSC)

The MSC is the "brains" of a wireless carrier's network. It monitors all active (powered on) mobile phones and all mobile calls throughout a system, and handles the switching of all calls as well. All cell base stations are connected to the MSC.

3.5.5   Interconnection to the Public Switched Telephone Network

In order for mobile calls to connect to the landline network, and for landline callers to reach a mobile subscriber, some type of linkage between the wireless network and the landline network must exist. These links are in the form of DS-*n* and OC-*n* level transmission systems between the wireless and landline network.

Each of these components of a wireless carrier network will be explored in detail in succeeding chapters.

3.6   Maps Used in Wireless System Design

The types of maps that are used in wireless system design are United States Geological Survey (USGS) maps. It is an FCC requirement that USGS maps be used for depicting service contours. Service (radio coverage) contours are a reflection of the 32-dBu signal measurement, which will be covered in another section.

All cell site locations are denoted using the universal positioning system of latitude and longitude, where latitude marks east/west and longitude marks north/south. Wireless engineers need to learn how to read latitude and longitude on maps regardless of the scale of the map. They need to learn how to read elevations on maps and the overall symbology. For example, they must determine where there are ridges, mountains, bodies of water, airports, and water towers.

The USGS maps that are used are 1:500,000 scale, where 1 in equals approximately 8 mi. In addition to large USGS maps, special maps known as "7.5-minute" maps are used for detailed analysis of topography, population centers, and roads. These maps are 1:24,000 scale, where 1 in equals 0.4 mi. These maps are used to determine terrain features in relation to where a cellular carrier wants to place a cell base station. The 7.5-minute map is a foundation for cell location searches.

Trivia: Some of the USGS maps haven't been updated since the 1920s. For example, southern Illinois USGS maps still date back to 1928. Obviously, a map that old and outdated would make it hard to determine a good cell location because the topology of the landscape would surely have changed in 70 years. New roads and buildings could have been constructed, and urban sprawl could have heavily affected the existing landscape.

USGS maps can be obtained through any of these sources: Rand McNally, the U.S. government, International Map Service, and similar companies.

3.7 POP Counts

POP counts is a term which means *population counts, or potential customers.* One person equals one POP count. The density of wireless markets is measured in terms of POP counts and *POPs covered,* and wireless markets are bought, sold, and traded on the basis of POP counts. Wireless carriers try to place the majority of their cell sites in areas where the population is dense. That way, the carriers can maximize the potential for generating revenue and obtaining return on the investment of building their systems.

3.8 AMPS Technical Specifications

AMPS cellular systems use FM radio transmission, where available spectral bandwidth is divided into 30-kHz channels, each channel capable of carrying one conversation or serial data stream at a time. FDMA (frequency-division multiple access) is the analog cellular modulation standard. FDMA describes the process of subdividing a large block of radio spectrum into many smaller blocks of spectrum. The 50 MHz of FCC-assigned radio spectrum is parceled into channels, and each channel is used for *either* the transmit or receive portion of a cellular telephone call. Cellular is a duplex mode of communications and two channels are needed for each call: one channel for transmit, one channel for receive. (See Chapter 5, "Radio-Frequency Channelization.")

Up until 1992, cellular radio had always used analog transmission. The radio communication between base stations and mobile phones transmitted an analog FM signal. Around 1992, digital radio was developed, initially using TDMA (time-division multiple access) technology (see Chapter 22, "Digital Wireless Technologies"). Since 1992, cellular carriers have deployed digital radio technology and service in order to increase the capacity of their systems.

AMPS cellular frequencies in the U.S. are:

Mobile transmit: 824—849 MHz (base receive)

Base transmit: 869—894 MHz (mobile receive)

See Figure 3-5 for a depiction of AMPS cellular frequency allocations.

Several fundamental attributes are needed for the realization of cellular service:

1. Frequency agility in the radiotelephone, which allows the mobile phone to operate on any given number of frequencies in the cellular radio spectrum (which subsequently allows for *call handoff*)

2. A contiguous arrangement of radio cells so that the mobile unit can always operate at acceptable radio signal levels

3. Call handoff capability

4. A fully integrated, transparent fixed network to manage these operations

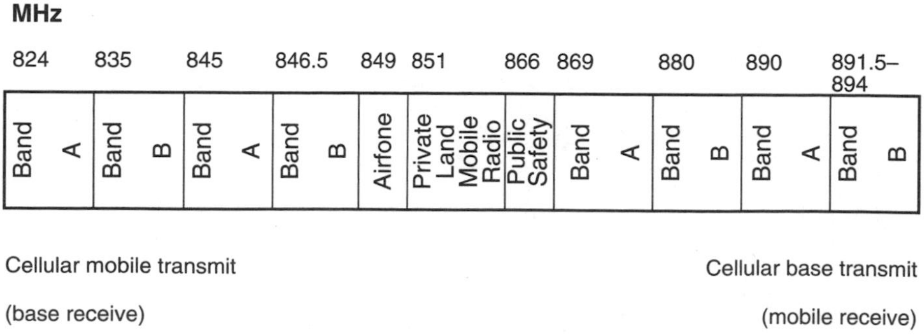

Figure 3-5
AMPS cellular frequency allocations.

3.9 Test Questions

True or False?

1. _____ The hexagon grid does not allow a cellular carrier to design its system to reduce or eliminate signal interference and facilitate long-term frequency planning.

Multiple Choice

2. The AMPS technical specification specifies:
   (a) FDMA as the modulation standard
   (b) AM (radio) transmission
   (c) FM (radio) transmission
   (d) Wave-division multiplexing
   (e) Short-wave radio transmission
   (f) a, b, and d
   (g) c and d only
   (h) a and c only
   (i) a and b only

3. Which technical concept sets cellular apart from all preceding mobile radio systems?
   (a) TDMA technology
   (b) Duplex functionality
   (c) Frequency reuse
   (d) None of the above

4. The frequency reuse plan is divided into cell groupings using how many cells, where the number of cells equals N?
   (a) 3
   (b) 10
   (c) 7
   (d) 416
   (e) 21

5. Which cellular engineering technical construct requires determining the geographic distance that is required between cells using *identical frequencies,* in order to avoid interference between the radio transmissions at these cells?
   (a) The Hertz theory
   (b) Call handoff
   (c) The distance-to-reuse ratio
   (d) The hexagon metaphor

The Cell Base Station

4.1   Overview

A cell (base station) is a physical region where a particular set of frequencies serves users, with adjacent cells using different frequencies to avoid interference. The cell base station serves as the *air interface* between the mobile phone and the cellular system; it is the first or last transmission leg of every cellular telephone call, whether that call is mobile-originated or mobile-terminated. Cell base stations are capable of handling multiple simultaneous conversations, and serve as the initial access point to the wireless network. There are multiple low-power radio transmitter/receivers (transceivers) located at each base station for the purpose of carrying customer conversations. Each cellular telephone is also a standalone transceiver. The cell base station is also known by the acronym BTS, for base transceiver station. The radio frequency emitted by a cell base station covers a geographic sector ("cell"), and cell base stations are created by carving up counties and/or cities into small areas called *cells*; hence the name *cellular.*

The term *cell* is derived from its hexagonal shape, which is similar to that of a cell within a bee's honeycomb. Cell sizes are not necessarily the same. Cells can range in size from less than a mile across, up to around 30 mi across, depending upon terrain, system capacity needs, and geographic location (urban or rural). Generally, cell sizes are very small in urban areas and larger in rural areas, as a result of population density and the overall amount of traffic on the wireless system.

Trivia: In Chicago in 1998, Cellular One's average cell size was 2 mi. In downtown Chicago, the cell size is half a mile or less. When the system was first built, the average size of a cell site was 5 to 10 mi. Cell sizes became smaller as traffic on the system grew and more cells were added to accommodate the increased capacity required. Cellular One and Ameritech Cellular each added about 250 cell sites in the Chicago area from 1992 to 1997.

4.2   Criteria and Methods for Cell Placement

The main criterion for determining where a cell base station will be located is where the customers are, or where they will be. This focus was initially on mobile subscribers (i.e., car phone users), and thus sites were usually placed along major roads and highways. Information on vehicular traffic is found from Rand McNally interstate road maps and from traffic studies by state and local departments of transportation. This information was used to determine the optimal placement for new base stations. The focus on where to install cell sites shifted in the 1990s, to include areas of high pedestrian traffic (i.e., shopping malls, downtown areas, nightlife areas). This was due to the immense popularity of the portable, pocket cellular telephone.

Overall, site selection today is based on population centers and densities, traffic densities on highways and major roads, the proximity of existing cells, and the existing frequency-reuse plan. Maintaining interference-free service while growing a cellular system is the most challenging task faced by wireless engineers.

4.3   Selecting Cell Site Locations

Several key parameters must be examined in determining an ideal location for a cell base station. First and foremost, what area do you want to cover with radio? What is the height of the terrain above mean sea level (AMSL)? Also, is the area in a flood plain? Ideal urban cell site locations are in business areas such as industrial parks and strip malls. Residential areas should be avoided when possible for aesthetic reasons (towers).

Cell sites are usually named after the area where they exist—for example, a town, a highway, or a mountain. Sometimes cell sites are named after company executives or employees, or are even simply numbered. The topology of the landscape sometimes makes it impossible to place cell base stations in convenient locations. For example, in mountainous regions such as the Western United States, some cells are placed right at the peak of a mountain in order to maximize the radio coverage area for that cell. These locations may be accessible only by helicopter and can be snowed in for months at a time. In order to reach some of these locations, a cell site technician might have to use a four-wheel drive vehicle for part of the run up a hill or mountain, then a snowmobile, and then walk the remaining distance to the site. These cells can run off of solar power or huge propane tanks that are filled before the winter weather sets in. The tanks can hold enough fuel to operate the cell nonstop until spring. In areas like those described above, some cells are even constructed on stilts above the projected snow line so that technicians can enter the cell unimpeded.

Trivia: Ameritech Cellular used a real *blimp* to simulate signal characteristics (and hence quality of coverage) at *potential* cell site locations. The blimp was about 14 ft tall and 40 ft long. It would be tethered at a potential cell location about 300 ft above ground level. There was a 10-dB antenna inside the blimp, with coax cable running down to ground level to an amplifier. By using the blimp to simulate a cell site, Ameritech engineers could determine whether the location would be good or less than ideal for construction of an actual cell. The blimp was an alternative to renting a crane, which would cost around $10,000 per test.

4.4   Cell Site Acquisition and Deployment

In new markets where cellular service had never been offered before, cellular engineers had to place cells in or near the population centers and along major roads and highways. Today, the locations for constructing new cell sites in existing cellular markets that have already been built-out can be a joint decision (a compromise) agreed upon by marketing and RF (radio frequency) design engineers. However, the proximity of existing cells and the existing frequency-reuse plan *must* be taken into account when examining potential cell placement options.

Wireless system engineers use *search rings* to designate areas where site acquisition specialists should seek land for a *new* cell site. Search rings describe a three-ringed geographic area that is deemed optimal for intended wireless radio coverage. The site acquisition specialists should try to obtain a new cell location within the center of the search ring's geographic area. If they are not successful in finding a location in the center of the area, then they should seek to place a new cell within the second ring in the target area, and possibly into the third ring, until they are successful. See Figure 4-1 for a depiction of the search ring concept.

Key: Where a new cell is ultimately placed will affect the frequency-reuse plan and RF power levels at nearby base stations.

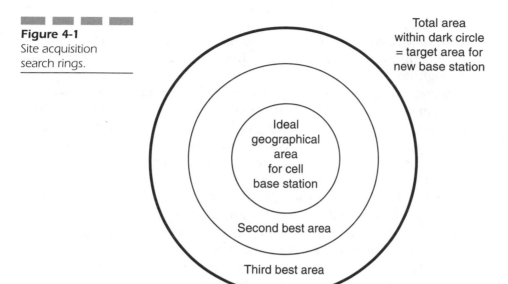

Figure 4-1
*Site acquisition
search rings.*

Cell sites may now be located at just about any land area or structure: on farms, in church steeples, on water towers, in hotel and motel signs, in apartments, and even on light towers at athletic fields. Some wireless carriers have even mounted their antennas on the top of grain elevators. When referring to cell site location in this manner, we are actually referring to places where the carrier's antennas may be mounted. Local municipalities are becoming much more stringent when approving new base station installations, especially in urban areas. This is due to community resistance to "unsightly" towers, and is sometimes known as the "NIMBY" phenomenon in municipal political circles—"not in my back yard." To that end, tower vendors have undertaken new initiatives to create towers that are camouflaged. (See Chapter 7, "Wireless Communication Systems Towers.") Many times wireless carriers have to dispatch special teams to testify before municipal zoning boards to get approval to install base stations/towers. These teams emphasize the benefits of wireless service to the community, and try to come to an agreement with the zoning boards as to what it will take to get the approval to install a new base station (i.e., tower) in a specific area. Sometimes zoning boards will even extract certain concessions from wireless carriers, such as mandating the carriers paint the towers a certain color, restricting the towers to certain heights, or mandating that the carrier share the site's tower with other wireless carriers. Some zoning boards will even request that the wireless carrier repave village parking lots or build new community parks, to demonstrate their "commitment" to the local community and "give something back" to the community for the right to install towers.

Today, many wireless carriers deploy their base stations on water towers, sometimes called water tanks. This type of deployment can be a win-win situation for all parties involved: the wireless carrier is spared from constructing an expensive tower, and a municipality receives a monthly payment from the wireless carrier for renting the space on the water tower. The monthly charge for renting space on top of a water tower can be anywhere from $500 to $3000, depending on the location of the water tower and the local government that is involved. Sometimes wireless carriers get very creative in their efforts to mount a base station on top of a water tower. One water tower in a suburb of Chicago allowed a wireless carrier to mount their base station (antennas) on the top of the water tower. However, the carrier had to repaint the water tower—with the name of the village on it—as a condition of being allowed to mount their site on the tower. On the plus side, though, the carrier also was allowed to paint their *name and logo* onto the tower as well, which resulted in free advertising! Sometimes water towers will even support multiple carriers (sets of antennas). One water tower in suburban Chicago has a base station (set of antennas) on *top* of the water tower, and *three* additional base stations (sets of antennas) from three different carriers on the *neck* or "stem" of the water tower. Sometimes antennas that are installed on a water tower have a "collar." A collar is a nylon or fiberglass wrap that is literally wound around the antenna mounting to make the antenna structure invisible to the general public.

In urban areas, many wireless carriers also deploy (new) base stations on rooftops of multistory buildings. Rooftops are ideal places for base stations as they represent a replacement for the construction of a tower; thus wireless carriers are spared from the capital cost of tower construction. The building owners also "win" in this situation as they are generating revenue in the form of monthly payments for space that might otherwise go unused.

Monthly rent for a rooftop base station can run anywhere from $500 to $3000, but the average cost is about $1000 per month. Similar to water tower rent, rooftop rental fees are dependent on the building owner and possibly the height of the building itself.

Another trend that has become commonplace regarding base station deployment is for municipalities to concentrate all wireless carrier cell sites (towers) within a small geographic area, around a half mile in diameter. These cell site "clusters" minimize the negative environmental impact of unsightly towers by containing all the sites to one specific area.

The industry average cost to build an entire base station can range from $500,000 to around $750,000, depending on the type and amount of equipment used and the manufacturer. This includes the costs for the land (lease or buy), the tower if one is necessary, the shelter that houses the equipment, the base station equipment (radio transceivers, etc.), and antennas and coaxial cable.

As recently as 1995, this cost was around $1,000,000. Many factors contributed to the decrease in the cost of cell sites, mainly the reduction in cell site equipment costs due to larger volumes being ordered by wireless carriers, specifically by the new PCS carriers. Components became smaller, too, hence cheaper.

The average lead time to install a new base station is approximately 6 months to a year. That reflects the time from "no lease" to a fully operational cell site.

4.5 Base Station Shelters

Older cell base stations, and some newer Personal Communication Services (PCS) base stations, deploy small self-contained shelters at the bottom of the towers at cell sites. The shelters house all the base station equipment that's necessary to make a cell site run. Shelters are also sometimes known by the acronym CEV, which stands for "controlled environmental vault." Newer base station manufacturers (i.e., Lucent Technologies) make self-contained *cabinets* that house all the equipment required to make a cell site function. These cabinets are stand-alone objects, and are 5 ft long, approximately 5 ft high, and 2 ft in depth. Cabinets have taken the place of the costlier shelter, for those carriers that choose to deploy them. The main purpose of shelters and cabinets is to keep equipment safe from the elements—cool and dry.

Today, with the explosion of wireless service and the accompanying proliferation of base stations, more communities are cracking down on "building and cabinet" pollution. They want shelters to blend in to the environment better. Shelters that are aesthetically pleasing are more sought after by community zoning boards. Just as disguised antennas and towers are gaining more recognition with zoning boards, attractive and less obtrusive shelters are on their way to setting a new trend in residential cell site planning. One option toward keeping shelters from being an eyesore is to keep them to a minimum. Today, a single tower can have several providers leasing space on it, but the drawback is that each of those providers will also be deploying their own shelter at the base of the tower. With the variety of shelter designs, colors, and fence heights, the base of a tower can begin to resemble a haphazard shanty-town. Modular shelters can help prevent this situation. These shelters are one physical unit that allows multiple carriers to be corralled together within a singular structure. Each carrier gets their separate space, contained behind a chain-link fence within the structure.

Another shelter option is structures that have brick or stone appearances on the outside. Brick shelters blend in to residential neighborhoods better than steel or fiberglass shelters. Sometimes a carrier will just paint a shelter, or add shrubbery to the outside, to have it match the surrounding buildings. This has proven to be an effective and low-cost method some shelter vendors have employed.

Some manufacturers even make bulletproof shelters, because in some areas the shelters are used for target practice. One shelter manufacturer has plans to make a shelter that looks like a *boulder*, so that it will completely blend into its landscape.

Security is also an issue to be remembered when installing a shelter. Keeping the equipment safe is the most important goal when selecting a shelter design, according to industry engineers. To that end, most shelters have 8- to 10-ft fences erected around them, topped with barbed wire. Shelter doors are made of steel, outfitted with steel pry-resistant latch guards and tamper-resistant hinge pins. One carrier provides motion detectors and alarms on some of its remote shelters to dissuade trespassers.

4.6 Cells on Wheels (COWs)

COW is an acronym which stands for *cell on wheels*. A COW is essentially a cell site (shelter) that is mounted on a flatbed tractor-trailer. It contains all the equipment that is normally used at a regular cell site. COWs are used for emergency purposes in any number of situations, for instance, if a tower fell down or an entire base station were inoperable.

COWs have built-in towers as well. The makeshift tower on a COW is usually a telescopic monopole that is erected from the truck itself. COWs *are* capable of providing for call handoff. Per industry slang, a large-sized COW is known as a bull; smaller-sized COWs are known as calves; and a large number of COWs is known as a herd.

MSCs can also be housed on tractor-trailers for emergency purposes. A perfect example is the story of a cellular carrier that owned the rights to operate in the Davenport, Iowa, MSA. During the Great Flood of 1992, this carrier's Davenport MSC was flooded and ruined. To solve the problem, the cellular carrier placed an MSC-on-wheels next to the site of the ruined MSC. When the carrier reconstructed the original MSC, it housed it on the second story of a building.

4.7 Test Questions

True or False?

1. _____ The cellular industry's average cost to build an entire cell site can range from $500,000 to $750,000.
2. _____ Cell sites can be located on towers only.
3. _____ Cellular radio transmissions are duplex transmissions.
4. _____ A cell site is usually named after the area where it exists.

Multiple Choice

5. The areas that system engineers designate for site acquisition specialists to seek land for a new cell site are known as:
   (a) Hex rings
   (b) "O" rings"
   (c) Search rings
   (d) D-R rings

6. Which cellular carrier used a blimp to simulate signal characteristics at potential base station sites (remember the year was 1983)?
   (a) Bell Atlantic Mobile
   (b) Southwestern Bell Wireless
   (c) Ameritech Mobile
   (d) Cell One
   (e) None of the above

Radio-Frequency Channelization

5.1 Paired Channels

There are a total of 832 channel *pairs* allocated per cellular market by the FCC. Since cellular is a duplex system like IMTS, two radio channels are needed for each cellular conversation. The channel transmitted from the base station to the subscriber's mobile phone is known as the *forward channel* or *downlink*. The channel transmitted from the mobile phone to the base station is known as the *reverse channel* or *uplink*. Paired channels are the combination of the forward channel and the reverse channel that are necessary for every cellular conversation to take place. The two respective frequencies of the mobile transmit (base receive) and the base transmit (mobile receive) are combined to form the *duplex channel* that is used for every wireless call. See Figure 5-1.

Key: Since the cellular industry is a duopoly, each wireless carrier in every cellular market is allocated 416 channel pairs. 416 channels are used for the base transmit/mobile receive side, and 416 channels are used for the mobile transmit/base receive side.

Most cellular carriers partition their 416 channel (pairs) into the $N = 7$ frequency-reuse format, the de facto industry standard. Other frequency-reuse plans exist, but they are not widely used. Base station transmit and

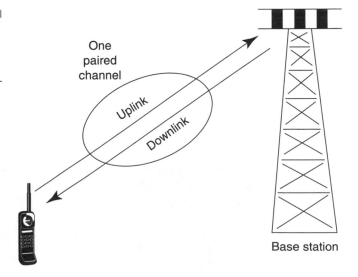

Figure 5-1
Representation of downlink and uplink, a paired channel.

receive bands are separated by 45 MHz of spectrum to avoid interference between cellular radio transmit and receive channels.

5.2   Channel Spacing

Channel spacing refers to the actual bandwidth space that is allocated for every cellular channel out of the total amount of cellular spectrum. In AMPS, the channel spacing is 30 kHz. Every uplink and downlink *each* occupies 30 kHz of bandwidth. This means that every cellular call actually occupies a total of 60 kHz of cellular spectrum: 30 kHz for the forward channel and 30 kHz for the reverse channel. In other words, 30 kHz for the uplink, 30 kHz for the downlink.

Each designated pair of frequencies—paired channel—that will be reused has been assigned a specific channel number under FCC guidelines and industry standards. This channel number equates *directly* to one specific paired channel, and its associated frequencies. See Table 5-1 for a *sample* of how paired channels are assigned in the cellular industry, using the 30-kHz channel spacing standard.

5.3   Control Channel

The control channel, also called the *paging channel*, is a data signaling channel that handles the administrative overhead of the cellular system via messaging between mobile phones, cell base stations, and the MSC. It is uscd to administer the following tasks:

TABLE 5-1

Sample of Paired Channels (A Band) and Their FCC Numbers

FCC Channel No.	Base Station Receive, MHz (Uplink)	Base Station Transmit, MHz (Downlink)	Mobile Transmit, MHz (Uplink)	Mobile Receive, MHz (Downlink)
1	825.030	870.030	825.030	870.030
2	825.060	870.060	825.060	870.060
3	825.090	870.090	825.090	870.090
4	825.120	870.120	825.120	870.120
5	825.150	870.150	825.150	870.150

- Setup of cellular calls, both mobile-originated and mobile-terminated.
- Locating (paging) cellular phones before connecting mobile-terminated calls.
- Collecting call information such as billing and traffic statistics.
- Autonomous mobile registration, i.e., registering phones on the system—both home and "roaming" phones. (See Chapter 19, "Cellular Call Processing.")

Of the 416 total channel pairs allocated per cellular carrier per market, 21 channel pairs are control channels. Like the 395 voice channels, the 21 control channels are also reused over and over again throughout cellular markets. All subscriber units—once they are powered on and throughout the time they are powered on—"tune" to the control channel in their assigned band (A or B band) from which they receive the strongest signal. Each subscriber unit automatically retunes the control channels in its band at predetermined intervals, based on system and carrier parameters. This interval can range from every 2 minutes to every 60 minutes. When a subscriber pushes the "send" button when placing a call, the phone again rescans for the strongest control channel signal.

5.4 Channel Sets

Each cell base station is assigned a particular number of cellular channels. This group of channels is known as a *channel set*. In an $N = 7$ frequency-reuse plan, there are 21 channel sets, with an average of 15 to 20 paired channels assigned per set. Channel sets are assigned on an alphanumeric basis. There are 21 channel sets because channel sets are assigned alphanumerically in groups of three, using the $N = 7$ reuse format. For example, there is channel set A1, A2, and A3. Then there is channel set B1, B2, and B3; and so on. This alphanumeric configuration was developed by AT&T (Bell Labs).

For example:

Channel Set Numbers* ($N = 7$)

1	2	3	4	5	6	7
A1	B1	C1	D1	E1	F1	G1
A2	B2	C2	D2	E2	F2	G2
A3	B3	C3	D3	E3	F3	G3

*Each alphanumeric designation equals one channel set.

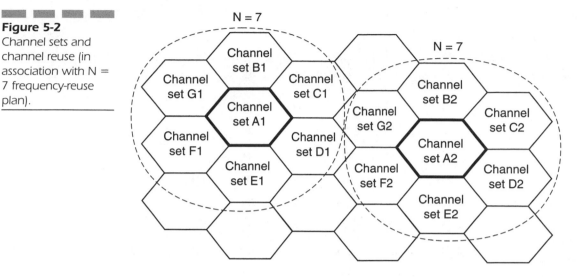

Figure 5-2
Channel sets and channel reuse (in association with N = 7 frequency-reuse plan).

Channel sets are assigned in this manner up to the letter G, because G is the seventh letter of the alphabet and this equates to the $N = 7$ frequency-reuse formula. Seven cell clusters times 3 channel sets equals 21 total channel sets. See Figure 5-2 for a depiction of 14 channel sets assigned in two $N = 7$ "clusters."

Key: The terms *frequency* and *channel* are synonymous. Frequency reuse is synonymous with the term *channel reuse* because all frequencies in the cellular spectrum have been divided into discrete channels.

See Table 5-2 for a *sample* of several standard channel sets, using FCC channel numbers to illustrate frequency assignments for each channel set.

TABLE 5-2

Sample Channel Sets (Plan A Non-Wireline)*

Signal (Control) Channels (One Control Channel per Channel Set)								
333	332	331	330	329	328	327	326	325
Group A1	Group B1	Group C1	Group D1	Group E1	Group F1	Group G1	Group A2	Group B2
312	311	310	309	308	307	306	305	304
291	290	289	288	287	286	285	284	283
270	269	268	267	266	265	264	263	262
249	248	247	246	245	244	243	242	241
228	227	226	225	224	223	222	221	220
207	206	205	204	203	202	201	200	199
186	185	184	183	182	181	180	179	178
165	164	163	162	161	160	159	158	157
144	143	142	141	140	139	138	137	136
123	122	121	120	119	118	117	116	115
102	101	100	99	98	97	96	95	94
81	80	79	78	77	76	75	74	73
60	59	58	57	56	55	54	53	52
39	38	37	36	35	34	33	32	31
18	17	16	15	14	13	12	11	10
Extended Channels (Expanded Spectrum)								
1020	1019	1018	1017	1016	1015	1014	1013	1012
919	918	917	916	915	914	913	912	911
711	710	709	708	707	706	705	704	703
690	689	688	687	686	685	684	683	682
669	668	667	666	665	664	663	662	661

*Each 3-digit number represents a channel pair.

▬▬▬ 5.5 Test Questions

True or False?

1. _____ Each carrier in every cellular market (A carrier/B carrier) is allotted 832 channel pairs.

2. _____ The radio transmission from the base station to the mobile phone is known as the *uplink,* or *reverse channel.*

3. _____ Channel spacing for the AMPS standard is 40 kHz.

4. _____ The terms *frequency reuse* and *channel reuse* are synonymous.

Multiple Choice

5. Cellular mobile transmit frequencies and cellular base transmit frequencies are separated by how much spectrum to ensure that interference is avoided?
   (a) 55 MHz
   (b) 110 MHz
   (c) 30 kHz
   (d) 45 MHz
   (e) None of the above

6. The control channel is a digital data signaling channel that performs which tasks?
   (a) Autonomous mobile registration
   (b) Locating (paging) cellular phones for connecting (terminating) land-to-mobile calls
   (c) Grooming of channel pairs
   (d) Collecting customer calling information such as billing and traffic statistics
   (e) Setup of all cellular calls
   (f) All of the above
   (g) a, b, d, and e only
   (h) c only

Radio-Frequency Propagation

6.1 Overview

Radio is an electromagnetic phenomenon in which energy travels in waves through a given medium. In the air, radio signals propagate at the speed of light: 186,282 miles per second. This equates to nearly a *billion feet per second*! The only significant functional difference between cellular telephone systems and the conventional landline telephone system is the *radio link* that connects the subscriber unit to the cellular network via the cell base station.

Radio frequencies differ in energy and ability to propagate media. Cellular frequencies are primarily reliant on direct waves; that is, they do not bounce off the earth's ionosphere as does short-wave radio. However, direct wave does not mean that cellular radio signals are limited to line of sight. Radio signals can bend through the atmosphere (*refraction*), bend around obstructions (*diffraction*), and bounce off obstructions and solid objects (*reflection*). Radio-frequency (RF) propagation refers to how well a radio signal radiates, or travels, into free space. In the wireless industry, it refers to how well RF radiates into the area where RF coverage must be provided, which is determined by RF engineers working for wireless carriers.

Omnidirectional ("all directions") radio-frequency propagation can be compared to the waves created by throwing a pebble into a pond. The waves made by the pebble emanate in all directions equally. The waves are proportional to the *size* and *weight* of the pebble and the *force* with which it was thrown into the pond. This pebble-in-a-pond analogy applies to the design of RF coverage for a cell. Actual RF coverage is mainly determined by three key factors: the height of the antennas (tower) at the base station, the type of antenna used at the cell base station, and the RF power level emitted. These three factors, together, equate to the size and weight of the pebble thrown in the pond, and the force with which the pebble was thrown.

6.2 Ducting

Ducting is defined as the atmospheric trapping of a cell base station's RF signal in the boundary area between two air masses, hot air over cold air or vice versa. Ducting of a cellular RF signal is caused by an atmospheric anomaly known as *temperature inversion*. Ducting is an anomaly of nature that can affect how well RF propagates through a given coverage area.

EXAMPLE   If the ground temperature is 30°F up to 500 ft, and then there is a layer
of ice cold or very hot air, the RF signal could be trapped between these
two air masses and propagate for as long as the duct exists. The duct
could go on for hundreds or even thousands of miles! This creates a
problem because instead of being absorbed by ground clutter (which is
usually desirable to a degree), once trapped the ducted signal could
cause interference in distant cellular systems.

Ducting is undesirable in RF propagation design, but also unavoid-
able. *Downtilting* of cellular antennas may compensate for ducting.
Downtilting is the act of electrically or mechanically directing the RF
emitted by the base station antenna toward the ground. (See Chapter 8,
"Antennas and Radio-Frequency Power.") The most common place for
ducting to occur is across large bodies of water like the Great Lakes.
Michigan base station transmissions are picked up routinely along the
Illinois and Wisconsin shore lines and Illinois stations are routinely
picked up in Michigan.

There are other physical anomalies that may affect propagation. What
may happen to the signal must be examined and hopefully compensated for by cellular RF engineers when designing cell sites. "Hopefully" is
the word because the anomalies mentioned above can have unpre-
dictable consequences on RF propagation.

6.3   Signal Fading

There are three basic types of fading of wireless signals: absorption, free-
space loss, and multipath fading, also known as *Rayleigh fading*.

6.3.1   Absorption

Absorption describes how a radio signal is *absorbed* by objects. When a
radio wave strikes materials, it can be absorbed. RF can be absorbed by
buildings, trees, or hills.

Key: Organic materials tend to absorb more RF signal than inorganic
materials. Pine needles are noted for absorbing a great deal of RF emis-
sions because their needle length is close to $1/4$ wavelength of cellular signals. Phasing of RF signals in the 800-MHz range lends itself to being
absorbed by pine needles because the length of the pine needle is equivalent to $1/4$ wavelength in the 800-MHz range.

Absorption can be compensated for by using higher-gain antennas and higher power levels, in order to cover the same geographic area. (See Chapter 8, "Antennas and Radio-Frequency Power.")

Absorption can also be compensated for by using shorter spacing between cells. The greater the amount of absorption of an RF signal, the less geographic area covered.

6.3.2　Free-Space Loss

Free-space loss, also known as "path loss," describes the attenuation of a radio signal over a given distance, or the path length of the signal.

Key: The higher the frequency of a radio signal, the greater the free-space loss.

Radio-frequency power levels determine at what point a signal will completely fade away to nothing.

6.3.3　Multipath Fading (Rayleigh Fading)

Rayleigh fading describes a condition where the transmitted radio signal is reflected by physical features and/or structures, creating multiple signal paths between the base station and the user terminal (mobile phone). When RF signals arrive at an antenna out of phase, they will either cancel each other out or supplement each other. One signal path arrives at an antenna (either mobile or base station) as a *direct* signal, while other signals are multipath, or *indirect,* signals. Indirect signals will reflect off any and all objects in the path between the transmitting antenna and the receiving antenna. These indirect signals arrive at receiving antennas via "reflection" paths.

These indirect signals can add to or subtract from the direct signal arriving at the antenna. The indirect signals *combine* with the direct signal to either complement or detract from the direct signal. This depends on whether or not the indirect signals are *in phase* or *out of phase* with the direct signal. If the indirect signals are in phase with the direct signal (i.e., "symmetric" wavelengths), then the indirect signals will *complement* the direct signal when they combine as they enter a receiver. If the indirect

■■■ ■■■ ■■■ ■■■

Figure 6-1
*Signal phasing. (a)
Signals that are in-
phase with each
other. (b) Signals that
are 180° out-of-
phase with each
other.*

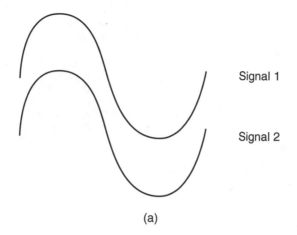

Signal 1

Signal 2

(a)

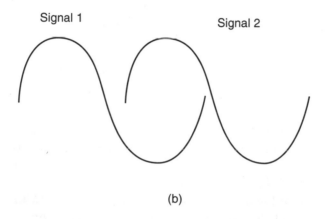

Signal 1

Signal 2

(b)

signals are out of phase with the direct signal (i.e., direct and indirect wavelengths are "offset" from each other), then the indirect signals will detract from—or weaken—the combined signal entering a receiver. See Figure 6-1 for an illustration of signals that are in phase and out of phase with each other. Multipath signals can reflect off bodies of water, vehicles, and even buildings on their way to a receiving antenna. See Figure 6-2.

6.4 The 800-MHz Band

The radio-frequency spectrum that was assigned to the cellular industry was mostly unused UHF television spectrum. The "800 Meg" spectrum

Figure 6-2
Multipath (Rayleigh)
fading.

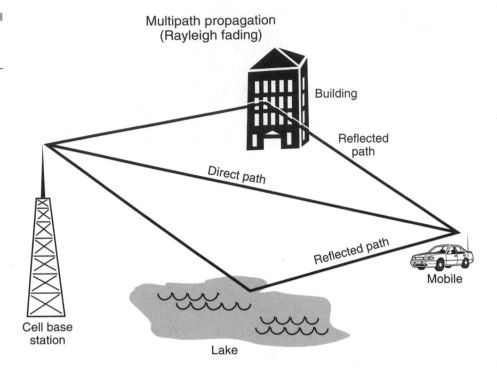

Multipath propagation
(Rayleigh fading)

Building

Reflected
path

Direct path

Reflected path

Mobile

Cell base
station

Lake

had been unused for years, and the TV broadcasters (networks) fought the FCC vehemently against reallocating their spectrum, even though it was grossly underutilized. The TV industry didn't use the spectrum, but also didn't want to give it up.

The following list describes the nature of radio propagation at 800 MHz and explains why the 800-MHz frequency range for cellular is mostly beneficial:

■ Very short signal wavelength (about 12 in).

■ Tends to be line of sight, similar to light itself. It is not really subject to skipping or bouncing off the ionosphere, as with short-wave radio.

■ Signal is easily reflected off buildings, cars, and trucks.

■ Signal is easily absorbed by foliage (i.e., trees, forest, etc.). This aspect of the 800-MHz spectrum can be both good and bad for cellular coverage. It is good because it allows for efficient frequency reuse. It is bad because it can cause problems when major coverage points of the cell (e.g., highways, towns) are in heavily wooded areas.

These factors, combined, make the 800-MHz spectrum ideal for cellular radio transmission.

6.4.1   In-Building Coverage

Today (2001), wireless carriers are still building out their "macrocell" network in their markets. This is the base station infrastructure in the outdoors, for example, rooftop cell sites, tower-based cell sites, and water-tank cell sites. From the carrier's perspective, any RF coverage that is afforded to customers in large (office) buildings is, to some degree, "incidental." Yes, the carriers would like all customers to be able to make calls regardless of where they are physically located, and this includes large buildings. In certain large buildings (i.e., convention centers, hotels, sports stadiums), wireless carriers *do* install small base stations known as "microcells." (See Chapter 12, "Enhancers and Microcells.")

Eventually, wireless carriers will specifically target large buildings for purposes of installing microcells to improve coverage within those structures. However, this activity will not take place until carriers believe that their macrocell networks are fully constructed. There are and will be certain impediments to expanding coverage into large buildings though.

Some large office buildings have a metallic coating added to their window glass to reduce the amount of heat taken through the window from direct sunlight. These are usually newer buildings, and their windows give them an appearance that resembles a green or gold mirror. Not only are the infrared rays of the sun blocked from these buildings, but radio waves are also severely attenuated and reflected in this scenario. This will severely impact in-building RF coverage from nearby base stations. In these instances, a wireless carrier could install an enhancer or a microcell in the building to compensate for the lack of coverage from macrocells in the "outer" wireless network.

6.5   Frequency Coordination

Frequency coordination is the effort to assign RF channels to base stations in such a way as to minimize interference within your *own* cellular system and *neighboring* systems of different wireless carriers. There are two kinds of frequency coordination: intramarket and intermarket.

6.5.1   Intramarket Frequency Coordination

This type of frequency coordination is internal to a cellular system (market). It is the effort to keep RF channel assignments as far apart as practical and feasible. It is based on a frequency-reuse growth plan using the hex grid. Internal frequency coordination is done to minimize interference and optimize cellular system operations.

6.5.2   Intermarket Frequency Coordination

This type of frequency coordination is external to a wireless carrier's system and involves coordinating frequency assignments with neighboring systems' border cells within 70 mi, according to FCC rule. The FCC dictates that all reasonable actions must be taken to limit and/or reduce interference between two different cellular systems. This usually relates to border cells between two markets. There are at least two ways to approach intermarket frequency coordination: (1) An RF engineer could simply call the engineering department of a neighboring wireless carrier. Engineers from both companies would exchange information regarding frequency assignments at border cells for their respective markets, and work together to resolve possible interference problems. (2) An outside vendor could be hired to handle the situation. This is not advisable, because it costs extra money, and the results may not be as accurate and reliable as approaching the other carrier directly.

6.6   System Interference

Interference is defined as any interaction of radio signals that causes noise or effectively cancels out *both* signals. Interference usually occurs between two transmitting radio signals whose frequencies are too close together, or even identical.

Base transmit and mobile transmit cellular frequencies were assigned with a 45-MHz separation between them to avoid interference. However, most interference experienced in cellular systems is still *internally* generated.

Higher frequencies may require less of a separation between transmit and receive than lower frequencies to reduce or eliminate interference.

6.6.1   Cochannel Interference

Cochannel interference occurs when there are two or more transmitters within a wireless system, or even a neighboring wireless system, that are transmitting on the *same* frequency (channel). Cochannel interference occurs when the same carrier frequency (base station) reaches the same receiver (mobile phone) from (two) separate transmitters. This type of interference is usually generated because channel sets have been assigned to two cells that are *not far enough apart* geographically; their signals are strong enough to cause interference to each other.

Key: Cochannel interference, when it occurs, is a by-product of the basic tenet of cellular system design: frequency reuse.

Though the basic principle of wireless system design is to reuse FCC-assigned frequencies over and over again throughout a system, it is also very important to ensure that the frequencies are reused far enough apart, geographically, to ensure that no interference occurs between identical frequencies (i.e., channel sets). Therefore, there must be enough base stations installed between cochannel cells to provide a level of protection to ensure that interference is thwarted and/or eliminated. In conjunction with ensuring that the cell sites are placed far enough apart geographically, the appropriate power levels must be maintained at cell base stations throughout a system as well, to avoid cochannel interference.

The following factors must be carefully determined to reduce the potential for cochannel interference:

- ■ RF power levels (most important factor)
- ■ Geographic distances between cochannel cells (cells using the same channel sets)
- ■ Types of antennas used

As more cells are added to a system operating on cochannel frequencies, it becomes more difficult to keep cochannel interference at a level that is not noticeable to subscribers of wireless service. Cochannel interference could manifest itself in the form of cross talk, static, or simply dropped calls.

Key: Cells that are cochannels must never be direct neighbors of each other. Cells with different channel sets must be placed in between cochannel cells to provide a level of protection against signal interference. Management of cochannel interference is the number one limiting factor in maximizing the capacity of a wireless system.

The following options are available to wireless carriers to reduce or eliminate cochannel interference:

1. Use downtilt antennas when and where appropriate (see Chapter 8)
2. Use reduced-gain antennas (see Chapter 8)
3. Decrease power output at base stations
4. Reduce the height of towers

6.6.2   Carrier-to-Interference Ratio

The carrier-to-interference ratio (C/I) is a measure of the desired signal the cell or mobile phone "sees" relative to interfering signals.

Key: Ideally, the goal in RF design is to have a C/I ratio of 18 dB or better throughout a wireless system to avoid cochannel interference. The 18-dB level was chosen by the cellular industry in the early 1980s to obtain a "clean, noise-free, landline-quality signal."

There should be an 18-dB difference between *any given cell* and all other cells (and mobile phones) throughout a wireless system. The carrier-to-interference ratio is analogous to the *signal-to-interference* ratio. The frequency-reuse plan is a tool used to keep the C/I ratio at the ideal level of 18 dB or better.

Key: As more cells are added to a wirelesss system, a migration occurs from a noise-limited system to an interference-limited system. The system has more potential to produce cochannel interference and/or adjacent channel interference (described below) as it grows.

6.6.3   Adjacent Channel Interference

Adjacent channel interference is caused by the *inability* of a mobile phone to filter out the signals (frequencies) of adjacent channels assigned to nearby cell sites (e.g., channel 361 in cell A, channel 362 in cell D, where cells A and D are in the same $N = 7$ frequency reuse cluster. Adjacent channel interference occurs more frequently in *small* cell clusters and *heavily used* cells.

Good system design can minimize adjacent channel interference temporarily by preventing adjacent channel assignments in cells that are near each other.

There are differing views as to why specifications for mobile phones are so poor concerning their inability to filter out adjacent frequencies. However, this situation will not be changed soon, so wireless carriers must live with the problem and try to engineer their systems better to avoid this type of interference.

6.6.4   Intermodulation Interference

There are other types of interference that occasionally plague wireless systems. The most common form of interference, other than cochannel and adjacent channel interference, is *intermodulation* interference (IM). If a cell site is colocated with other radio-based services, *intermodulation interference* may result. Competent engineering practices should overcome this interference.

Intermodulation interference describes the effect of several signals mixing together to produce an unwanted signal, or even no signal at all. Another type of IM is created by mobile phones themselves. If a customer is on a call in close proximity to a cell site operating on the opposite band (A band/B band), the power from all of the radio channels in the cell can cause the receiver in the mobile phone to overload. When this happens, the result will be a dropped call. This problem is a direct result of mobile phone manufacturers' reducing the cost of producing the phones. To resolve this situation, all wireless carriers operating in a market may have to place cell sites near each other so that a stronger signal is maintained in the mobile phone. Then, the mobile phone will not overload and drop the call.

6.7   Test Questions

True or False?

1. _____ Downtilting of cellular base station antennas may compensate for the ducting phenomenon.

2. _____ There are three main types of signal fading: absorption, free-space loss, and cross-channel equalization.

3. _____ Inorganic materials tend to absorb more RF signals than organic materials.

4. _____ Ducting of a cellular RF signal is caused by an atmospheric anomaly known as *temperature infraction*.

Multiple Choice

5. The three key parameters that determine actual RF coverage from any given cell base station are:
   (a) RF power levels at the cell base station
   (b) The height of the cellular base station antenna
   (c) The type of tower used
   (d) The type of antenna used
   (e) How many buildings are in the area
   (f) None of the above
   (g) a, b, and d

6. The higher the frequency of a radio signal, the greater the:
   (a) Absorption rate
   (b) Free-space loss
   (c) Antenna reverberation
   (d) Cochannel interference
   (e) None of the above

7. Cochannel interference describes which of the following?
   (a) Interference between mobile phones
   (b) The inability of the mobile phone to filter out signals of contiguous cellular channels (e.g., channel 331, channel 332)
   (c) Interference between two base stations transmitting on the same frequency
   (d) All of the above
   (e) a and b only

8. The *multipath* effect of cellular RF signals bouncing off of many objects, resulting in signals arriving at a mobile antenna at different times, in or out of phase with the direct signal, is known as:
   (a) Fessenden fading
   (b) Uplink fading
   (c) Rayleigh fading
   (d) Free-space loss
   (e) Ducting
   (f) None of the above

9. Radio signals in free space travel at:
   (a) The speed of sound
   (b) Twice the speed of sound
   (c) Half the speed of light
   (d) The speed of light
   (e) None of the above

Wireless Communication System Towers

7.1 Overview

According to the FCC, there were 500,000 antenna structures nationwide as of January 1995. With the buildout of PCS services since 1996, this number has probably increased significantly.

There are three basic types of tower available for erection at a cell site: monopole towers, free-standing towers (also known as self-supporting or "lattice" towers), and guyed towers. The type of tower that is actually installed at any given cell base station may be dictated by company policies, operational needs (such as a minimum tower height), or local zoning restrictions. Each choice in tower design has advantages and disadvantages that must be weighed against its intended use.

Tower design and construction methods are predicated on the tower's purpose, location, average weather conditions at the site, projected load, and future expansion needs. With proper research and commonsense planning, a properly constructed and maintained tower can last for decades. How much space a tower structure will occupy is determined by the type of tower, its maximum height, and its support design.

Key: In the wireless industry, towers may be used not only for mounting base station antennas, but also for mounting microwave antennas. The microwave antennas (dishes) are used in a wireless carrier's fixed network. (See Chapter 16, "The Fixed Network and System Connectivity.") The practice of deploying extensive microwave radio systems for the fixed network usually depends on the particular wireless carrier.

Local municipalities are becoming much more stringent in approving tower locations. In some communities (usually urban areas), wireless carriers sometimes meet with very stiff resistance when attempting to install cell base stations with "unsightly" towers.

When building cell base stations, the goal when determining tower heights is to get above the ground clutter (trees and buildings). Tower height depends on how many cell sites a carrier installs overall, taking into account frequency reuse. The minimum height depends on the area in question (i.e., the overall height of the ground clutter) and the extent of the territory you want to cover.

7.2 Site Survey

Many of the factors that are used as input in a site survey are simply common sense. When conducting a site survey for a communications tower, a wireless carrier includes all the applicable criteria that were decided upon concerning cell placement in general. (See Section 4.2, "Criteria and Methods for Cell Placement.") For example, the carrier takes into account the geographic area to be covered and whether land for the cell base station can be obtained at a reasonable price (lease or own). The site for any type of communications tower should be level and, *ideally,* easily accessible by a semi–tractor trailer and a heavy-duty crane.

After a location is selected for a cell base station, a contractor is chosen to erect the tower. The next step in the tower construction process is the evaluation of the proposed location. It is the responsibility of the party erecting the tower to research deed restrictions and zoning regulations in order to avoid conflict with neighboring property owners and any local and state agencies.

The tower site should also be free of overhead obstructions such as power lines and trees, and the foundation area should be void of utility lines or pipes. The preferred site should undergo a comprehensive soil analysis to determine whether the ground will adequately support the structure and what type of grounding system will be required to meet grounding standards. Soil testing should be conducted by a soil engineer who is familiar with tower and foundation construction. Testing procedures should encompass multiple geological core samples from across the site. Detailed soil analysis will weed out undesirable locations such as all-rock terrain, swampy or sandy areas, or even old, plowed-under farm fields. Soil testing should also be done to determine if there is any toxic waste on the site. If the site is purchased by a wireless carrier and toxic waste is found afterward, the new owner (the carrier) must pay to have the waste cleaned up.

Trivia: The site selection project manager for one wireless carrier did *not* do thorough soil testing at a proposed location for a cell site before he purchased the land. Once excavation for the cement pads for the shelter and tower had begun, it was discovered that the site was a plowed-over onion field. During the remainder of the construction phase, the workers had to wear gas masks. A much deeper hole also had to be excavated for the cement pads, resulting in a much higher overall cost to build that particular cell site.

When necessary, helicopters or large tracked vehicles have been used to transport cell site buildings and towers to remote site locations.

7.3 Monopole Towers

Monopole towers are constructed of tapered tubes that fit together symmetrically, and are simply stacked one section on top of another. Older versions of the monopole tower (and some newer versions) use 40-ft tube sections. The base of the monopole is bolted onto concrete pads, and no support cables are required. Tower construction companies use giant augers around 8 ft in diameter to drill deep into the ground to create the hole that concrete is poured into, which forms the base of the tower. The bottom of the monopole tower is then bolted to this concrete base. Depending on soil conditions at any given location, the hole that is drilled could be anywhere from 8 to 40 ft deep. Monopole towers are predominantly located in urban areas, and are usually installed for aesthetic purposes. These towers are usually not more than 150 ft tall, but can reach heights of 250 ft. Of all the tower types, they have the smallest physical footprint from a size perspective. See Figure 7-1.

7.3.1 Monopole Tower Advantages

Monopole towers are the most aesthetic of the available types of tower and require the least amount of land area for installation. Of the three tower types, these towers are the easiest to get through zoning hearings because of their less "offensive" appearance.

Trivia: Some municipalities have been known to insist that wireless carriers paint monopole towers a sky-blue color. The thinking is that the tower will then blend in with the sky. In reality, this strategy may make the tower stand out even more starkly against the landscape and the horizon.

To overcome possible opposition of communities to tower erection, some manufacturers offer monopole towers that are disguised as trees, complete with fake trunks, fake branches, and fake leaves. Monopole towers can also be disguised as light poles (parking lot lighting). These towers are much more expensive than a normal monopole tower. To date, pine and palm tree towers have been designed and marketed. However,

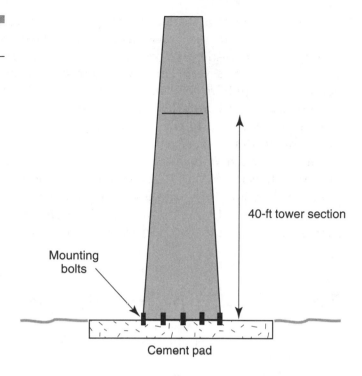

Figure 7-1
Monopole tower.

40-ft tower section

Mounting
bolts

Cement pad

although these types of towers make it a little easier for wireless carriers to gain community acceptance of tower construction, there are times when they are still out of place. There is an instance of a wireless carrier that installed a fake-tree monopole in the middle of a forest. However, the average tree height of the forest was 100 ft, while the fake-tree tower was 140 ft tall. Thus, the fake-tree tower still looked out of place, and of course it never dropped one leaf in autumn.

7.3.2   Monopole Tower Disadvantages

The main disadvantage of this type of tower is that, once its required height has been determined, and the tower has been ordered and erected, sometimes the only way to change the height of the antennas on the tower is to pull the tower out and erect a completely new tower from scratch. This is a relevant issue because it sometimes becomes necessary to raise or lower cellular antennas because of changes in the wireless system's design, namely the frequency-reuse plan. Remember, the height of base station antennas is one of the three key factors that dictate the scope of coverage at any given cell. New antennas and microwave dishes *can* be fastened to the sides of monopole towers—they are welded on with brackets. This practice is not easy and it is very expensive, but it is sometimes necessary because monopole towers are tapered, which prevents simply raising or lowering the antenna platform to a new position. Erecting a new monopole tower just to change the height of the antennas is also an expensive option.

Some carriers have even been known to actually cut off the tops of monopole towers and install new mounting structures for the antennas. However, there are three reasons why this activity is not a good idea: it is expensive, it could void the warranty on the tower, and it is risky because it could affect the structural integrity of the pole.

7.4   Free-Standing Towers

The free-standing tower, also known as a self-supporting or lattice tower, is the median tower type from a cost and structural perspective.

Free-standing towers are *three-* or *four-sided* structures, constructed of steel cross-arm sections that increase in diameter as they approach the ground. Like an elongated pyramid, free-standing towers are widest at the base and narrowest at the top. No cabling is used to support these towers; they are supported by having their bases anchored in cement with large bolts. Free-standing towers may be installed almost anywhere. They are frequently seen next to major highways in urban areas (e.g., toll plazas). See Figure 7-2.

7.4.1   Free-Standing Tower Advantages

Free-standing towers require minimal land area for installation (more land than monopoles, but less land than guyed towers).

If base station or microwave antennas need to be moved higher or lower because of changes in wireless system design, this can be accomplished with minimal effort on a free-standing (or guyed) tower simply by moving the antennas or dishes up or down the tower structures.

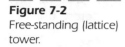

Figure 7-2
Free-standing (lattice)
tower.

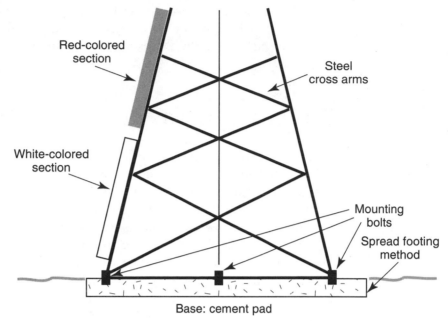

Base: cement pad

7.4.2 Free-Standing Tower Disadvantages

Free-standing towers are usually not more than 300 ft high. They become *extremely* expensive at heights of 150 ft or greater. This is because the higher they are, the more foundation they require, and the more substantial the structural elements must be. They also require a more extensive anchoring system. This involves more excavation work, which can become very costly.

7.4.3 Anchoring of Free-Standing Towers

There are two methods by which free-standing towers can be anchored:

- The *spread footing method* involves pouring a large single slab of concrete, usually about 5 ft deep, along the entire diameter of the base of the tower. The tower structure is then anchored to the slab of concrete with bolts. This method of footing is built to withstand a great amount of twist and sway due to wind conditions. It is also the preferred method for anchoring free-standing towers.

■ The *pier footing method* involves boring deep holes in the ground, one for each leg of the tower, and filling them with concrete. Bolts fasten each leg of the tower into the concrete.

7.5 Guyed Towers

Guyed towers are constructed of identical triangular steel cross arms, usually in 20-ft sections, that are supported or "guyed" by tensioned support cables held in place by concrete anchors on the ground. Their cost is linear with their height. The higher the tower, the more expensive it becomes. Guyed towers are the tallest tower type, and can reach heights up to 2000 ft. They are more predominant in rural areas, because of fewer zoning restrictions. Guyed towers for wireless carriers operating in RSAs are usually built to heights of around 300 ft.

Guyed towers are the same width at the base and the top. They are supported by thick guy cables that anchor the tower in place from *three* sides. The guy cables are placed at designated points throughout the height of the tower, these points being dictated by the physics of tower design (overall tower height) and tower loading factors (see Section 7.7, "Tower Loading"). See Figure 7-3.

Figure 7-3
Guyed tower.

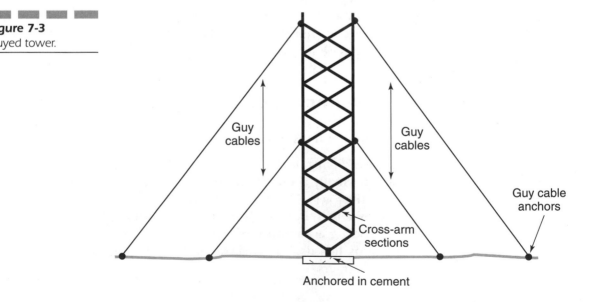

7.5.1   Guyed Tower Advantages

Guyed towers are the cheapest of all tower types to install. The cost of guyed towers is incremental with their height. A wireless carrier seeking to purchase a site for a guyed tower location can make a deal with the land owner to purchase *only* the small guy anchor plots of land plus the tower's base area. When this is done on farms, fences are placed around all the guy anchor plots to prevent farm machinery from damaging the guy anchors and the tower itself.

7.5.2   Guyed Tower Disadvantages

The main drawback of guyed towers is that this tower type requires more land than other tower types because it is necessary to install the guyed support cables to hold the tower up. Cabling for guyed towers can require a radius that is 80% of the tower height! The guy cables extend beyond the base of the tower for up to 200 ft in *three* directions. It is possible that guy cables can cause interference to RF antennas that are not mounted on top of the tower, but there is no way to model this.

Guyed towers also require more maintenance because guy-wire tension needs to be checked, and possibly adjusted, at periodic intervals.

The installation of a guyed tower can become costly if the wireless carrier cannot strike a deal to buy just the land necessary for the base of the tower plus the guy anchor plots. Because guyed towers require more land, when the wireless carrier must buy the entire plot of land necessary to construct the guyed tower, the land purchase alone can become a major cost.

7.6   Structural Design Options for Free-Standing Towers

When ordering any type of tower, it is important for wireless carriers to specify the wind loading projected for the tower—the maximum wind speed the carrier estimates the tower will encounter at a given location. The carrier also needs to specify the number and size of antennas and the number and size of microwave dishes to be mounted on the tower. These factors will determine the size and thickness of the steel for towers, and for guyed towers, it will determine the total number of guy wires that will be required to support the structure.

Wireless carriers can choose among solid-leg, tubular-leg, and angular-leg tower designs. These three design options usually apply to free-standing towers.

7.6.1   Solid-Leg Towers

Solid-leg towers are those towers whose tubular structural elements are solid steel. Solid-leg towers withstand damaging conditions better than other designs. Towers built solid from mill-certified steel are more resistant to sway, wind, rust, and freezing than hollow-leg designs. A solid-leg tower will sway less in high winds, reducing the potential for damage to the antennas or the tower itself, and it is also less vulnerable to damage that may result from antenna installation. This tower type can be built in various sizes.

With solid-leg towers, there are no parts that can collect moisture and subject the structure to ice-induced cracking or splitting. Corrosion is less likely to occur because manufacturing standards require 25% more galvanizing on solid-bar legs than on tubular legs.

7.6.2   Tubular-Leg Towers

Tubular-leg towers are those towers whose structural circular crossbar elements are *hollow.*

These towers, like solid-leg towers, can also be manufactured in assorted sizes. However, the tubular-leg tower's hollow structure can collect moisture, risking ice-related damage. Rust and corrosion can also occur internally from condensation or pooling of water. Therefore, effective drainage must be monitored and maintained. Drainage holes ("weep" holes) at the bottom of towers must be periodically checked to ensure there is no blockage, and that no rust is evident. Ice can and will break a tubular tower if it is allowed to form extensively within the tower structure.

Construction costs for this type of tower are similar to those for solid-leg towers, but repair bills are more expensive. If the thickness of the tubes is substantial enough, tubular-leg towers work best with light loads in locations where no expansion of tower load is expected.

7.6.3　Angular-Leg Towers

Angular-leg towers are those towers whose structural elements are solid L-shaped steel beams. The components of these towers are not tube-based. The use of angular-leg towers depends on the size and thickness of the steel of the structure. In other words, the size and thickness of the steel may dictate the tower loading factor (i.e., how many antennas and dishes may be placed on the tower, whether coax cable or waveguide cabling is used, etc.).

With angular towers, freezing is not an issue because of their solid nature. Of the three structural models, angular-leg design is the most expensive to build. Depending on the thickness of the steel, it may also be the most susceptible to damage sustained from swaying, because the thin flat surfaces may be less sturdy than solid circular crossbars. In the year 2000, approximately 90% of *guyed* towers that are installed are angular-leg towers.

7.7　Tower Loading

Tower loading defines the weight and wind force a tower is expected to withstand. The amount of support a tower will need is dictated by projected tower loading factors. Tower loading is based on:

- The tower's height.
- The total number of antennas to be mounted on the tower and whether they will be base station antennas or point-to-point antennas such as microwave dishes, or both.
- Existing weather patterns at the tower location.
- Maximum wind load (velocity) expected at the tower's location. Most towers are designed to sustain maximum wind loading of around 100 mi/h.
- Amount and thickness of coaxial cabling attached to the tower.

7.8　FAA Regulations

It is the responsibility of the Federal Aviation Administration (FAA) to regulate objects that project into or use navigable airspace. Radio towers, by FCC rules, must abide by FAA regulations. The FCC is the enforcer for the FAA with relation to tower construction.

7.8.1   The 7460-1 Form

The 7460-1 FAA form must be filed to show tower site locations for towers higher than 200 ft or for towers that are in the glide path of any airport. The form must show the site's latitude and longitude, the base elevation of the site, and the overall height of the tower and antennas so the FAA can determine if the tower is a hazard to air navigation. Latitudinal and longitudinal coordinates are designated using three sets of numbers, the first set designating "degrees," the second set designating "minutes," and the third set designating "seconds."

Wireless carriers must also file a 7.5-minute USGS map of the cell site with 7460-1 forms delivered to the FAA. The FAA will study the application and notify the wireless carrier if the tower must be lit and/or marked.

It takes the FAA about 1 month to process a 7460-1 form. Once the FAA receives the 7460-1 form, it will notify the carrier if a 7460-2 form is also required.

7.8.2   The 7460-2 Form

The 7460-2 form is used to inform the FAA when a carrier will begin and end construction of a tower, so the FAA can notify area airports. Carriers may need to obtain *special clearance* by the FAA before constructing towers, depending on their response to the filing of 7460-1 and -2 forms.

7.9   Tower Safety

As a rule, any objects taller than 200 ft are required by the FAA to have some type of marking and lighting. The various lighting and painting requirements are established by the FAA, but enforced by the FCC. Tower marking is optional, depending on the type of lighting used. Marked towers have alternating red and white sections painted on the tower structure. Typical lighting options include red incandescent lights, white strobe lights, or a combination of both known as *dual lighting.* Of the two types of tower lighting, strobe lights are used in the daytime and incandescent red blinking lights are used at night. The FAA allows carriers to use both types of lighting if they choose (dual lighting).

Trivia: Wireless carriers can be fined if the paint on marked towers has faded too much. The FAA used to warn carriers it was preparing to come out to inspect the quality of the coloring (paint job) of the towers, but the FAA now comes with no warning. Some carriers have swatches of the proper quality of the coloring for tower markings, and occasionally use the color swatches to ensure they are within the legal scope of how brightly the tower should be marked.

All tower lighting controls should have alarms to indicate lamp failures (for the red incandescent lights), flash failures (for white strobe lights), or power outages. If the lighting fails for any reason, the FAA must be quickly notified so aircraft operators can be forewarned. In most cases, if a safety light on a tower goes out, a site manager has 15 minutes to write a report on the defective light or risk a large FAA fine for endangering the flying public.

Because lighting is so critical, it is monitored throughout the construction process. Lights must be installed at two intermediate levels as construction of the tower progresses, and at the top of the tower. This is to ensure that the highest point on the tower is visible to aircraft at all times. A completed tower will have one beacon at the top, and may have two or more at midlevel, depending on the overall height of the structure.

The FAA is very serious about enforcing tower lighting and placement. One wireless carrier failed to properly light a tower during the construction phase, causing a medical evacuation helicopter to crash into the tower. Everyone aboard was killed. The wireless carrier was fined millions of dollars for this infraction. Another wireless carrier erected its tower higher than stated on the 7460-1 form it filed and was fined millions of dollars because of the proximity of the tower to an airport.

7.10   Tower Leasing

It is always an option for wireless carriers to lease space on an existing tower by approaching the owner of the tower. Wireless carriers generally try to lease space whenever possible, to save the cost of building their own tower. Leasing tower space will also appease local zoning boards who are developing a negative attitude toward construction of new towers in their municipalities.

But leasing tower space may or may not be less expensive than erecting your own tower, depending on many factors. How long do you want to lease the tower space? Would it prove to be cheaper to build your own tower, depending on how long you plan on keeping your license in a particular market?

In areas in the western part of the United States, Bureau of Land Management (BLM) land is extremely hard to buy or lease because the federal government owns it. Some wireless carriers examine whether other tower owners have prime real estate where no other options are available. Sometimes wireless carriers end up leasing tower space to their in-market competitors. This has become a prominent issue with the advent of PCS carriers building out their markets. Many municipalities today even *require* carriers to colocate wherever possible, to avoid having additional towers constructed in their towns.

The tower leasing industry is growing rapidly. In the late 1990s, a tower consolidation movement has taken place, where it has become the norm for multiple carriers to share one tower. Along these lines, there are a number of companies that do nothing but build or buy towers and lease the space to wireless companies. This includes cellular, PCS, Enhanced Specialized Mobile Radio (E/SMR), as well as paging carriers.

Some wireless carriers have sold their towers to management companies, offering space for other wireless operators to build out on. Since the Telecommunications Act of 1996 was signed, many companies have become innovative when it comes to leasing tower space. One company that began buying up radio stations across the country also saw the potential to gain revenue by leasing space on all the towers that came with the radio stations they purchased. Since so many wireless carriers have faced challenges from communities blocking additional tower construction, this company saw a market need and developed the idea. In each market, this company had 4 to 8 towers that were usually 300 to 500 ft tall. Each of those towers offered opportunities not only for the company that owned the towers but also for cellular and PCS carriers as well.

7.11 Tower Maintenance

7.11.1 Tower Inspections

Cellular carriers, who own some of the oldest towers in the industry, say the secret to a long-lasting tower is a strong maintenance program. Since the early 1980s, carrier towers have held up fine. But extra weight from ice, wind, other carriers' equipment, and additional antennas will weaken any tower over a period of time. If carriers don't catch the wear and tear early on, it will eventually put a tower completely out of commission. A strict maintenance program can ensure that towers hold up for many years. Regular inspections, ideally by companies who are not the original builders of the tower, every 1 to 3 years are the key. An important practice for any tower inspection is to review the original tower drawings. The purpose of having tower drawings is to ensure factual information about the age, structure, and capacity of a tower. Even if a carrier is leasing tower space from another company or carrier, they should request the tower drawings or inquire about tower maintenance records. Many tower inspections are done visually. An inspector usually will check ground-level supports, bolts, and paint degradation, but will not climb the tower. However, the best way to see what kind of maintenance a tower needs is to get out there and climb it. If the tower is not visually inspected from the top down, corrosion at the top of the tower and eroding metal interiors will often go unnoticed.

Guyed towers require inspections more regularly, mainly due to stretching of the guy cables. For these towers, the basic items that should be checked during any tower inspection are the guy wires, cable tension, and the connections of the guy cables. Vandals, inappropriate installation, and weather are all determining factors in cable maintenance.

For all tower types, carriers should also check the foundation and the condition of the foundation bolts. There are ways of checking a tower's foundation with detailed accuracy too. Stress-wave technology can ascertain the depth of a foundation, giving a carrier a better sense of its load limit without drilling into the foundation itself.

7.11.2   Weather, Corrosion, and Loading

Towers take constant hits from Mother Nature, especially in coastal locations or areas susceptible to tornadoes and storms. Carriers may want to inspect a tower after a harsh storm, because the tower may have been exposed to wind speeds higher than the tower's design load can handle. To help prevent corrosion in these areas, carriers should avoid using hollow leg towers because it may be difficult to inspect the inside of these structures to verify that corrosion is not occurring inside the tower. Because towers can rust from the inside out, carriers should consider conducting an ultrasound test to determine if there is any interior corrosion. But other measures can be taken to protect towers from corrosion. One major carrier frequently repaints their towers in regions with intense sunlight, heat, or wind to maintain sufficient protection of the tower structure. Ice puts the most significant load on a tower, so carriers operating in the northeastern United States engineer their towers and tower foundations to handle the extra weight that ice exudes on a tower.

Weather is not the only thing that can weaken towers. Equipment overloading causes most tower failures. Failures start in the legs of lattice towers, and monopoles deteriorate because of buckling or compression failure in the steel. If the steel is painted, not overstressed, and inspected on a routine basis, a tower can last for decades with good maintenance. Towers should also be reinforced if a carrier is adding new equipment or when inspections report damage to tower "members" (legs).

7.11.3   Preventive Maintenance

As with any equipment, preventive maintenance can be the key to keeping a tower in good shape over a long period of time. There are four key areas that carriers should focus on, to maintain a long-lasting tower.

1. Towers should be painted on a regular basis to keep water from corroding them. Sealants can be used around joints to ensure protection against the elements. Towers should never be painted in the summertime because when the temperature goes up, the tower's metal expands, causing paint to shrink and crack when the metal contracts during cooler temperatures.

2. Whether a carrier rents space on a tower or owns its own structures, coax cable needs to be maintained too. Coax should be changed out every 10 to 15 years. The thicker the cable, the longer it will last. Thinner cables should be checked on a more regular basis. Coax cables can also be the victim of target-practicing vandals. If a bullet grazes coax cable, it could result in a significant signal loss over time

3. Carriers should send their engineers and installers to tower maintenance certification programs, which are frequently offered by tower and consulting companies.

4. Tower lights should be replaced regularly if any of them appear weak (red incandescent) or slow (strobes).

▬ ▬ 7.12 Test Questions

True or False?

1. _____ One of the advantages of monopole towers is that they are the most aesthetic of all the tower types.

2. _____ Failure to notify the FAA of a lighting failure at a tower can result in stiff fines.

Multiple Choice

3. Guyed towers:
   (a) Are the tallest tower type
   (b) Are more predominant in rural areas
   (c) Are the same width at the base and the top
   (d) Are the most expensive tower type to install
   (e) a, c, and d only
   (f) a, b, and c only

4. Which structural design for towers allows for rust and corrosion to occur internally from condensation or pooling of water?
   (a) Solid-leg towers
   (b) Angular-leg towers
   (c) Tubular-leg towers
   (d) None of the above

5. Tower *loading* describes which of the following?
   (a) How many cross-arm sections a guyed tower can support
   (b) How deep the cement footing is placed at the base of a tower
   (c) How much wind resistance a tower can withstand
   (d) How much weight a tower can support
   (e) All of the above
   (f) c and d only

Antennas and Radio-Frequency Power

8.1   Overview

An antenna's function is twofold. To *transmit,* an antenna must take the radio signal that is applied to it and broadcast that signal as efficiently as possible. An antenna transforms alternating electrical current into a radio signal that is propagated through the atmosphere. Conversely, to *receive,* an antenna transforms a radio signal into an alternating electrical current that is decipherable to the receiving equipment. Put simply, antennas radiate energy into space, and collect radio energy from space.

There are two main types of antennas used in the cellular industry. *All* antennas fall under one of these two categories: omnidirectional or directional antennas. There are a multitude of omnidirectional and directional antennas available today.

The word *decibel* is used to compare one power level to another, and is denoted by the symbol dB. Antenna propagation is measured in decibels. For example, a 9-dB antenna is commonly used at omni cells. Cellular base station antennas range in average length from 2 to 14 ft, and weigh from 10 to 50 lb.

In some cases there are areas known as "antenna farms." An *antenna farm* is a plot of land that has been dedicated by the owner to the placement of towers for all different types of wireless services, offered by all types of wireless carriers (e.g., cellular, PCS, paging, E/SMR).

8.2   Omnidirectional Antennas

An omnidirectional (omni) antenna is one that radiates an RF signal equally in *all* directions (360°). Cells using omnidirectional antennas are known as *omni cells.* Omnidirectional antennas are sometimes referred to as "sticks" because of their appearance. See Figure 8-1.

8.3   Gain

The term *gain* refers to how the radiation pattern emitted from an antenna is reshaped.

Figure 8-1
Omni antenna
propagation.

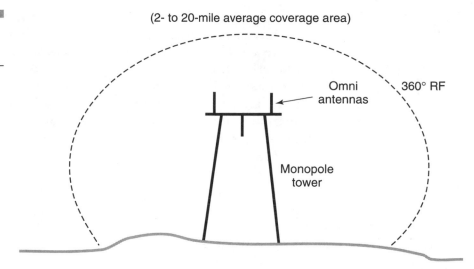

(2- to 20-mile average coverage area)

Omni
antennas

360° RF

Monopole
tower

Key: Gain is a measure of how much an antenna's vertical radiation pattern is compressed to the horizontal plane. This compression results in pushing the RF propagation outward *horizontally* from a radiating antenna, at a given radio power setting. How far the RF propagates outward horizontally is equivalent to how far the RF coverage extends from any given base station.

Gain is a logarithmic progression. It is represented by the width of a radiation pattern, or how much the pattern has been electrically compressed (vertically) and expanded horizontally.

Analogy: Think of the RF coverage area of a cell as being the surface of a balloon. Gain is achieved (and increased) if the balloon is compressed on the top and the bottom simultaneously. As the balloon is squeezed, it flattens out and widens. This flattening represents gain on the horizontal plane. With omnidirectional antennas that radiate in a 360° pattern, the simplest way to achieve gain is by stacking multiple $1/4$-wavelength antennas one on top of the other, and feeding the energy emitted by these antennas together in such a way as to add each antenna's vertical radiation pattern together in phase. When the patterns of all the antennas are added in phase, their signals complement each other instead of canceling each other out, thus contributing to the total, composite signal.

The more radiating $1/4$-wave antennas that are stacked on top of each other (while their patterns are kept in phase), the more gain is produced.

When we speak of stacking multiple $1/4$-wavelength antennas one on top of the other, this is actually a reference to *installing* antennas that were *designed and built* to a specific gain level. This type of antenna configuration is known as a collinear array, described below.

In the balloon analogy, as the amount of downward force is increased, the balloon becomes flatter, but its circumference is increased. The amount of "bulge" represents gain. As the gain is increased, the vertical pattern becomes progressively more compressed, until at 12 dB the vertical radiation pattern is only about 4" in height (or thickness) at the antenna. The average gain for an omnidirectional antenna is 9 dB. The higher the stated decibel level is for a given antenna, the higher the gain that the antenna will deliver.

8.4 Collinear Array Antennas

A collinear array antenna is a predominant type of omnidirectional and directional antenna. To produce a collinear array antenna, one or more $1/4$-wave antennas are stacked inside the antenna *radome*. The radome is the white or gray fiberglass shell that encases antennas. It appears tubular, and is what we all see and think of as the actual antenna. Each antenna that is stacked inside the radome represents an incremental increase in gain, in most cases in increments of 3 dB. However, these incremental increases in gain are not without some cost. Each time 3 dB of gain is produced, the overall length of the antenna is doubled. If a 3-dB-gain omni antenna is 3 ft in length, a 6-dB-gain antenna will be 6 ft in length. A 9-dB-gain antenna would then be 12 ft long, and a 12-dB-gain antenna would be 24 ft long. See Figure 8-2. The antenna elements that comprise the collinear array antenna will be longer for antennas (antenna elements) that are used for lower frequencies. This length in the antenna elements correlates directly to the fact that lower frequencies emit longer radio wavelengths. Thus, antennas used for lower frequencies will be physically longer than antennas used for high frequencies, if the gain assigned to both of those antennas is the same. Put differently, if a carrier deploys a 9-dB antenna that operates at 500 MHz, and a 9-dB antenna that operates at 800 MHz, then the 800-MHz antenna will be physically shorter due to the shorter wavelength inherent with RF emitted at 800 MHz, the higher frequency.

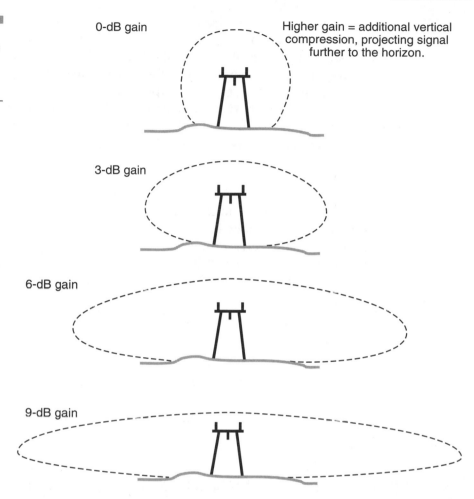

Figure 8-2
Omni gain (antenna propagation) for a collinear array antenna.

0-dB gain

Higher gain = additional vertical compression, projecting signal further to the horizon.

3-dB gain

6-dB gain

9-dB gain

8.5 Downtilt Antennas

With a very high tower and a high-gain antenna, *coverage shadows* may be created near the tower. To compensate for coverage shadows, downtilt antennas were developed specifically for the wireless industry by antenna manufacturers. A downtilt antenna is an antenna whose radiation pattern can be tilted a specified number of degrees downward. The downtilt antenna decreases distance coverage horizontally, but increases signal coverage closer to the cell site. A common place to install a downtilt antenna is at a cell site that's on a very tall tower or a hill.

Downtilt antennas are also used to compensate for what is known as the *far-field effect* in wireless systems. The far-field effect occurs when the radio coverage projected from site A may completely overwhelm the intended coverage area of site B. Site A may transmit and receive into site B, theoretically leaving site B unused. This would not only be terribly inefficient, but would be a terrible waste of equipment and frequency resources at cell site B. The deployment of a downtilt antenna at cell site A would ensure that the intended radio coverage from site A stays within its designated coverage boundary. The far-field effect may occur in areas where a cluster of sites has many towers of different heights.

Key: Mechanical downtilting of antennas can be done only with *directional* antennas. Downtilting of omnidirectional antennas is done *electrically,* not mechanically. It is not feasible to mechanically downtilt an omni antenna. Electrical downtilting is accomplished by adjusting the phasing of the RF signal that is fed to the collinear antenna elements. A wireless carrier would purchase a downtilt antenna from a manufacturer. The carrier does not mechanically manipulate the elements of a "regular" antenna to convert it into a downtilt antenna.

Downtilt antennas can shrink omni cell coverage areas while maintaining static RF power levels. Downtilt antennas improve portable (phone) coverage. They can also be used to pull coverage away from a market border while maintaining static power levels. See Figure 8-3.

8.6   Criteria Used to Determine Selection of Base Station Antennas

Base station antennas are much more sophisticated and utilize a much wider variety of designs than mobile phone antennas. One reason for this difference is that base station antennas are required to have a higher gain, ordinarily between 6 and 12 dB for omnidirectional antennas. In some cases, 0-dB-gain omni antennas have been used. This scenario might involve engineering for a microcell.

The type of base station antenna that is chosen in any situation depends on many factors:

- The size of the area to be covered.
- Neighboring cell site's configurations.

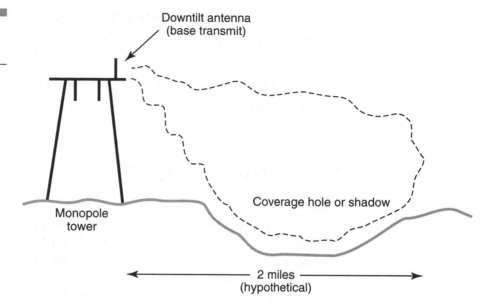

Figure 8-3
Downtilt antenna
function.

- Whether the antenna is omnidirectional or directional.
- If it is a directional antenna, the antenna's beamwidth.
- How much of the allotted RF spectrum the antenna can utilize. In the cellular industry, carriers can get antennas for the transmit band only, the receive band only, or both the transmit and receive bands.

8.7   Mobile Antennas

Many mobile antennas are collinear array antennas. Mobile collinear array antennas consist of two antennas that are stacked together, separated by a "pigtail" in the middle. The purpose of the pigtail is to act as a phasing coil. The pigtail puts the two separate antennas in phase, so that the antennas *add* their respective signals together to produce gain, instead of canceling each other out. Mobile antennas are the weak link in the wireless system because they are the one element over which wireless carriers have the least amount of control.

Mobile antennas are usually mounted in less than optimal places on cars. Ideally, they should be mounted in the center of the car roof to obtain the best signal strength. However, they are mounted at many other places on cars, and sometimes, when users go through a car wash, a glass-mount antenna gets bent at a 45° angle, inhibiting optimal signal quality. Car-mounted antennas are also exposed to harsh elements such as ice, snow, and road salt, which also detract from their optimal functioning.

Most mobile antennas deliver 3 dB of gain. However, there are some models that deliver 5 dB of gain.

8.8 Antenna Quality

Quality varies with both base station and mobile antennas used in the wireless industry. Some antennas can stand up to the elements, while some do not. Some will leak, some will literally fall apart. Ultraviolet rays and salt are especially harsh on mobile antennas. Antenna quality is measured in several ways. One way is to determine how many dissimilar metals are used in the antenna. This is important because the more different metals that are used, the more likely the antenna will develop corroded junctions, which can cause intermodulation (IM), or signal mixing. For instance, two signals may go in and four signals may come out. Ultimately, this results in a degradation of signal quality.

8.9 Radio-Frequency (RF) Power

RF power is defined as the amount of radio-frequency energy, in watts, delivered by the base station radio to the base station's transmit antenna. RF amplifiers used at the cell site determine power and the amount of RF energy that is delivered to the base station antenna by the base station radios (transceivers). There will also be loss of the signal as it propagates through the coaxial cable from the transceiver up to the transmit antenna, due to *impedance*. This is factored into RF design by wireless system engineers.

8.9.1 Effective Radiated Power (ERP)

ERP stands for *effective radiated power.* It is determined by the gain of the antenna (mobile or base station) times the power delivered to the base of the antenna. ERP is measured in watts.

Key: It is important to remember this rule of thumb: When the power output of amplifiers (radios) is *doubled* (*in watts*), a 3-dB increase in gain is achieved.

Example: Ten watts of radio energy directed into a 10-dB-gain (base station) antenna equals 100 W of ERP.

Example: A carrier is using an 8-W power amplifier (PA), with a 15-dB-gain antenna and 3 dB of line loss (impedance). This equates to a net gain of 12 dB. The number 8 is the factor in this equation, because that is the power delivered by the amplifier to the antenna. In this scenario, with the 8-W amplifier, presume we start with 0-dB gain. Three decibels of net gain would deliver 16 W ERP (8-W amplifier with 3 dB of gain equals a doubling of ERP). Six decibels of net gain would equate to 32 W ERP. Again, every additional 3 dB of gain equates to a doubling of ERP. Nine decibels of net gain would equate to 64 W ERP. Twelve decibels of net gain would equate to 128 W ERP. So, since we know we have a net gain of 12 dB, the ERP for a base station using the equipment listed above would be 128 W. Again, notice that each time we increase net gain by 3 dB the ERP doubles.

Author's note: Actual formulas used to determine ERP can be very granular, as they represent logarithmic relationships. That information is beyond the scope of this text.

8.9.2   Allowable Power Levels

8.9.2.1   Cell Base Stations   In the cellular industry, carriers are allowed to use up to a *maximum* of 500 W ERP at cell base stations. The range of 100 to 500 W is primarily used in RSA base stations. BTS power levels used depend on the frequency-reuse pattern. Ideally, the maximum power level needed to provide coverage is used, but no more than that.

The average power level for MSA base stations ("macrocells") is 20 to 100 W to cover an area from 8 to 30 mi. Again, this level will depend on terrain in the coverage area and antenna gain. The overall range of ERP in a wireless system can be anywhere from $\frac{1}{10}$ to 500 W ERP.

Base stations can and will throttle down the power of a mobile phone if the mobile phone is emitting too much power, for whatever reason. This will avoid overloading equipment at the base station. One example of this is when a mobile phone is located too close to the base station. The BTS will send a message over the control channel telling the mobile phone to throttle down its power to a specified level.

TABLE 8-1

Mobile Station

Power Level	Class 1	Class 2	Class 3
0	4.000	1.600	0.600
1	1.600	1.600	0.600
2	0.600	0.600	0.600
3	0.250	0.250	0.250
4	0.100	0.100	0.100
5	0.040	0.040	0.040
6	0.016	0.016	0.016
7	0.006	0.006	0.006

8.9.2.2 Mobile Telephones Mobile phones have allowable power classes assigned to them by the FCC. These power levels have a range from 0.006 W (six hundredths of a watt) to 4 W. Mobile phones (auto installations) have higher allowable power levels—they emit more power because they run off of car batteries.

Table 8-1 depicts allowable power levels for different types of mobile phones in a cellular AMPS system, in *watts.*

8.10 Test Questions

True or False?

1. _____ The radome is the white or gray fiberglass tube that encases antennas.

2. _____ Gain is not a logarithmic progression.

3. _____ When the wattage output of the amplifiers/radios is *doubled,* a 10-dB increase in gain is achieved.

Multiple Choice

4. Omnidirectional gain is defined as the compression of RF at which plane?
   (a) Horizontal
   (b) Bidirectional

(c) Directional

(d) Vertical

5. When wattage output of amplifiers/radios is doubled, how much of an increase in gain does this represent?

(a) 10 dB

(b) 5 dB

(c) 25 dB

(d) 500 dB

(e) 3 dB

6. What term is used to compare one radio power level to another?

(a) Hertz

(b) EMI

(c) Decibel

(d) Channelization

(e) None of the above

Base Station
Equipment
and
Radio-Frequency
Signal Flow

9.1 Tower Mounting of Omnidirectional Transmit and Receive Antennas

When looking at a base station tower, a casual observer would notice two stick-shaped antennas pointing downward, and one antenna pointing upward. If so, this means that this is an omnidirectional cell base station.

9.1.1 Receive Antennas

The two antennas that are pointing downward on the base station tower are receive antennas (RX 0 and RX 1). These antennas *receive* RF signals at the base station from the mobile telephone. They receive the uplink signal—the mobile-to-base signal. The diversity receive antenna (the second antenna) compensates for Rayleigh fading in the uplink to the base station. Diversity is a tool used to *optimize* the signal received by a base station antenna. It counteracts the negative effects of Rayleigh fading.

The two receive antennas provide for *space diversity,* as follows:

- When a wireless customer presses the send button on the phone to place a call, the signal enters *both* of the receive antennas, and travels down the coax cable on the tower into the base station.

- At that point, a device known as a *comparator* in the base station transceiver examines *both* of the received signals, and selects the best signal of the two received. The comparator *continues* to dynamically select the best of the two receive signals for the duration of the wireless call. The call can change from being carried by one receive antenna to being carried by the other receive antenna dozens or hundreds of times during an average wireless call.

Key: Space diversity can provide anywhere from a 0.5- to 12-dB difference in received signal strength between one receive antenna (RX 0) and the other receive antenna (RX 1).

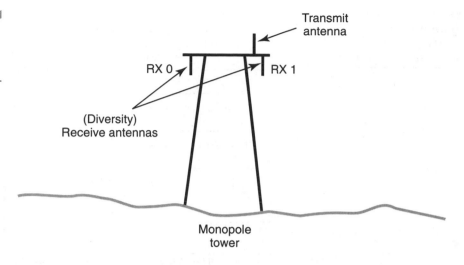

Figure 9-1
Tower mounting of omnidirectional antennas (side view).

9.1.2   Transmit Antennas

The antenna that is pointing upward at an omni cell base station tower is the transmit antenna. This antenna is used to transmit from the cell base station to the mobile phone. These antennas transmit the down-link signal—the base-to-mobile signal. See Figure 9-1.

In omnidirectional cells, the reason that the receive antennas point downward and the transmit antenna points upward is to reduce or eliminate intermodulation interference. Even though the cellular transmit and receive bands are separated by 45 MHz, having the omni transmit and receive antennas mounted so they point in different directions helps to reduce the potential for interference.

9.2   Cell Site Configuration*

9.2.1   Transceivers (Base Station Radios)

Transceivers, or base station radios, have two receive ports (diverse receive signals) one transmit port, an "audio-in" channel, an "audio-out" channel, and a data line. The data line runs the functions of the transceiver as

---

*For illustration purposes only, the equipment and configurations described in this section reflect an "average" Nortel-supplied analog cell site, where each cell site handles an average of 16 cellular paired channels (radios/transceivers). One of these 16 channels is a control channel, since each cell has a control channel assigned to it.

■■■■ ■■■■ ■■■■ ■■■■
Figure 9-2
Typical base station
transceiver.

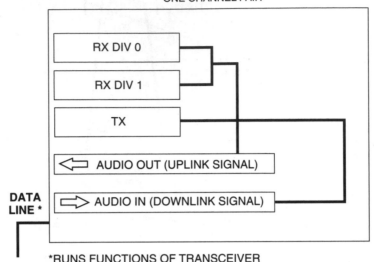

BASE STATION
TRANSCEIVER

ONE CHANNEL PAIR

RX DIV 0

RX DIV 1

TX

⟵ AUDIO OUT (UPLINK SIGNAL)

**DATA
LINE *** ⟹ AUDIO IN (DOWNLINK SIGNAL)

*RUNS FUNCTIONS OF TRANSCEIVER

the transceiver communicates to the base station and the MSC. See Figure 9-2 for a depiction of a typical base station transceiver.

9.2.2 RF Signal Flow through a Cell Site: The Downlink

For illustration purposes, the scenario of a mobile-terminated call will be used here. A wireless customer is receiving a call. From the PSTN trunk that was assigned by the MSC, the incoming call is routed to the (land) "audio-in" input into a *transceiver.*

Key: The MSC has already assigned a paired channel and frequencies to this call. The assigned channel and frequencies will remain the same for the duration of the call, or until a call handoff is executed.

A power amplifier is used in the downlink signal path to boost the radio signal up to 45 W maximum output. Amplifier signal levels are measured in decibels. There is one power amplifier assigned to each channel pair, or transceiver.

If a given cell site has a 16-channel radio bay, then there will be four 4-channel combiners at that cell site. The purpose of the combiner is to allow for the use of one antenna for multiple radio channels, rather than having one separate antenna dedicated to each channel at the cell. In other words, without a combiner, there would theoretically be 16 antennas at the cell. Combiners eliminate the cost and poor aesthetics of having to install a separate antenna and run coax cable down the tower for each and every radio channel. A by-product of the combiner's call processing is that it reduces the signal level by half.

Key: A decrease in a power level by one-half means that the overall power has been cut by 3 dB. Therefore, the signal power at the star connector is now *half* what it was when it entered the combiner: It is now $22\frac{1}{2}$ W.

Each combiner port must be tuned to the frequency of the radio channels assigned to the cell by a wireless technician.

A star connector links together all four of the 4-channel combiners.

The RF signal is now routed into a duplexer. The function of the duplexer is to enable both the transmit and receive signals to be routed through the same antenna. The call is then transmitted out from the cell base station to the mobile telephone at a base station transmit frequency. Duplexers are not always used by all carriers, but they avoid the cost of having to use two receive antennas instead of one.

Key: The entire process described above goes hand in hand with the actual processing of a wireless call. (See Chapter 19, "Cellular Call Processing.")

Figure 9-3 gives a graphical depiction of signal flow over the downlink.

9.2.3 RF Signal Flow through a Cell Site: The Uplink

Once an uplink signal is received into the RX 0 and RX 1 antennas at the cell site, it is transmitted down into the base station into a special bandpass *filter.* The function of this equipment is to filter out all frequencies except the receive frequency for the specific channel processing a given wireless call.

Figure 9-3
Downlink signal pro-
cessing in a typical
Nortel AMPS cell site
configuration.

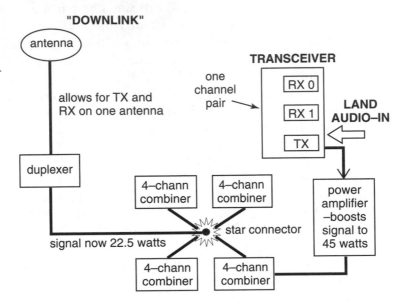

RF SIGNAL FLOW THROUGH A CELL SITE:

The signal is then routed to a *low-noise amplifier,* or "preamp." The purpose of the preamp is to boost the received signal to a level that's strong
enough to be split into 16 outputs. This is necessary because of loss
from impedance as the signal flows through the coax cable. The low-
noise amplifier contributes very little noise to the received signal.

The call is then routed to a device called a *multicoupler.* The function of
the multicoupler is to split the two received signals independently into 16
coaxial receive outputs, which are then routed into 16 separate transceivers.

Key: Each transceiver has two receivers: one connected to the RX 0
diversity antenna and one connected to the RX 1 diversity antenna.

The multicoupler's function is somewhat analogous to the combiner's
function in that both pieces of equipment allow for the consolidation
and interleaving of signals.

Both received signals now enter the transceiver. The transceiver constantly selects the best of the two received diversity signals via the comparator, and continues to process the call. Once the transceiver has selected the best receive signal, it then routes the call to the "audio-out"
output, and then to the MSC. Once the call reaches the MSC, the switch

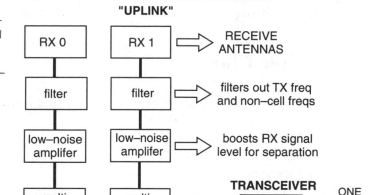

Figure 9-4
Uplink signal process-
ing in a typical Nortel
AMPS cell site config-
uration.

completes processing. Least cost routing is taking place at this point. (See Chapter 18, "Wireless Interconnection to the Public Switched Telephone Network.")

See Figure 9-4 for a graphical depiction of signal flow over the uplink.

9.3 Test Questions

1. Draw a diagram that shows the RF signal flow over the downlink into the base station. Include each piece of cell site equipment in your diagram, along with a brief explanation of the function of the equipment.

True or False?

2. _____ A device in cellular base station transceivers known as a *complexitor* examines both of the received signals in the RX 0 and

RX 1 antennas, and constantly reselects the best signal of the two received during the processing of cellular telephone calls.

Multiple Choice

3. A by-product of the combiner's call processing is that it:
   (a) Increases signal level by 25%
   (b) Decreases signal level by 33%
   (c) Decreases signal level by 50%
   (d) Increases signal level by 50%
   (e) None of the above

4. The concept that describes the use of two antennas to receive a signal, allowing for the best signal to be chosen for processing, is known as:
   (a) Time diversity
   (b) Space duplexity
   (c) Space diversity
   (d) Space-time redundancy
   (e) None of the above

5. Which piece of cell site equipment splits the two received signals from the RX 0 and RX 1 antennas into (an average of 15) independent coaxial receive outputs?
   (a) The multicoupler
   (b) The combiner
   (c) The transducer
   (d) The star connector
   (e) None of the above

10

Wireless System Capacity Engineering

The capacity of a wireless system is a function of the number of radios per cell in each market. There are many approaches that a wireless carrier can employ to manage system capacity. In this chapter we review the predominant options used to address expanded capacity in wireless systems. What methods any particular carrier uses can depend on many factors. First, capacity-engineering practices could be carrier-specific. In other words, the engineering management at certain carriers could favor certain practices or technologies. Second, cellular carriers may be more inclined to use certain engineering practices that a PCS carrier would not use. For instance, since cellular carriers still have some AMPS base stations, they may be more inclined to split those cells or use directed retry as a stop-gap measure to obtain more capacity. On the other hand, a PCS carrier may be more inclined to employ smart antenna technology, especially if that carrier's digital standard is CDMA. Generally speaking, some of the practices listed below may be seldom used today by wireless carriers. Ultimately, the type of practices or technology that a wireless carrier will use depends on the carrier and the nature of their network.

10.1   Cell Splitting

Cell splitting is the process of dividing two existing cells into three cells. A new cell is installed in an area that is around the shared border of the two existing base stations. Cell splitting is done to increase system capacity so the system can handle more traffic in a specific area.

Key: Cell splits are dictated when cells in a given area are consistently at 80% of capacity.

The existing frequency-reuse pattern (i.e., channel set assignment) must be closely adhered to when doing cell splits, in order to reduce or eliminate interference. The power levels of all surrounding base stations affected by a split must be adjusted downward to reduce expansive coverage overlap that can cause interference. Since splitting adds cells to a system, the intended coverage area of all nearby cells will shrink; cells will become smaller. Today, most carriers try strenuously to avoid cell splits due to the vast amount of work associated with constructing and deploying a brand new base station, which could cost upwards of

Figure 10-1
Cell splits.

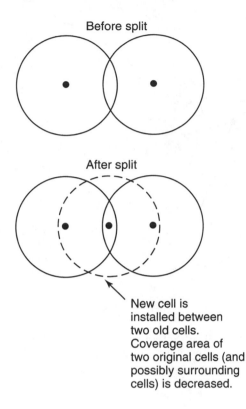

Before split

After split

New cell is
installed between
two old cells.
Coverage area of
two original cells (and
possibly surrounding
cells) is decreased.

around $500,000 (average). Most carriers will try to use other technologies to gain more trunking efficiencies from their existing cell base stations. See Figure 10-1 for a depiction of cell splits.

10.2   Overlay/Underlay Engineering

Overlay/underlay describes the process of (figuratively) building a small cell *inside* a larger cell. This is accomplished by tuning radios (channels) at adjacent cell sites so that the dB (power) levels of every other channel within the cell are different (about 5- to 10-dB difference). This provides for the reduction and/or elimination of adjacent channel interference between neighboring cell sites. Roughly half of the channels in the base station will have lower power assigned to them, and the other half will have higher power assigned to them. The theory holds that adjacent

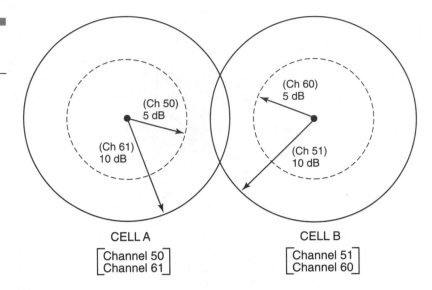

Figure 10-2
Overlay/underlay
concept.

channels could be assigned to the lower power, "interior" channels in one base station, and the neighboring base station adjacent channels would have higher power assigned to the "exterior" channels. See Figure 10-2 for an illustration of the overlay/underlay concept.

Key: Overlays are implemented to reduce (adjacent channel) interference. Indirectly, implementation of overlays will ultimately augment system capacity by allowing more flexible use of adjacent channels in neighboring cells. See Figure 10-2.

10.3 Directed Retry

Directed retry is the process by which a given cell (cell A) reroutes its call origination attempts to an adjacent cell (cell B) or other nearby cells for processing. Directed retry is utilized if there are no channels available to process a call in cell A. In some cases, a long list of cells could be on the retry list to ensure that a call origination attempt goes through. These cells are usually also neighbor cells used for call hand-off purposes. This may account for the occurrence of long hang times (postdial delay) when wireless customers are attempting to place calls. Obviously, since calls generate revenues for the carrier, the carriers will

Figure 10-3
Directed retry
process. Mobile must
be detectable by con-
trol channels at each
base station.

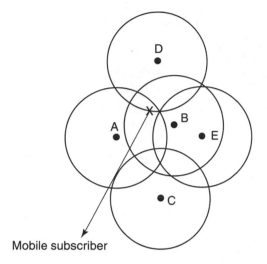

Mobile subscriber

develop and use all means necessary to ensure that all call origination attempts reach the system for processing. *Note:* In order for any cell or cells to provide for directed retry, the control channels at these cells *must* be able to detect the call origination attempt. These cells must be able to receive the signal from the originator's wireless phone. The retry process is coordinated by the MSC, and can be activated and deactivated on a cell-by-cell basis.

The directed retry feature may be only a temporary measure to allow customer's access to the system, until the capacity of cells in the affected area can be increased. See Figure 10-3.

10.4   Sectorization

10.4.1   Directional Antennas

A directional antenna is an antenna that shapes and projects a beam of radio energy in a specific direction, or receives radio energy only from a specific direction, employing various horizontal beamwidths. There are many types of directional antennas used by wireless carriers: log periodic, Yagi, phased-array, and panel antennas. One application of directional antennas may be to keep undesired radio signals out of specific areas, such as a neighboring cellular market. Directional antennas are mostly used in sectorized cells, as described below.

10.4.2 Sectorization Overview

As more cochannel cells are added to a cellular system, it becomes necessary to migrate from an omni antenna configuration at cell base stations and begin to "sectorize" some of the cells in a wireless market. Sectorization is a complex process that essentially subdivides one omnidirectional cell into three to six separate subcells, all colocated on one tower, using multiple sets of directional antennas. Sectorization helps to minimize or eliminate cochannel interference and optimizes the frequency-reuse plan. To sectorize a cell, an equilateral platform that resembles a triangle is mounted onto the tower. Each side of the platform is called a *face*, with three directional antennas installed per face. The directional antennas propagate the different frequencies assigned to each sector (or "face").

Key: Most wireless carriers, both cellular and PCS, usually deploy only three sectors per cell site.

When an omni cell is sectorized into three sectors, antenna faces are installed at discrete intervals of 120°. Each sector has its own assignment of radio channels. Each sector also has its own assigned control channels, and will hand calls off to its adjacent sectors.

Key: From a graphical viewpoint, sectorization takes a regular circle (representing an omni cell) and converts it into a three-section pie chart, all sections being equal. See Figure 10-4.

10.4.3 The Grid Angle ("Azimuth")

When implementing sectorization into a wireless system and/or sectorizing omni base stations, wireless carriers develop and adhere to what is known as the "grid angle," also known as the "azimuth" of the system. The grid angle is the orientation, in degrees, of the first face of all sectorized cells from true north. The grid angle allows for all base stations throughout a sectorized system to be laid out *symmetrically* in order to avoid interference between cochannel sectors. In other words, a wireless carrier in city X would have the antenna platform faces at all of their base stations oriented in the exact same way, at the exact same angle, throughout the area.

Figure 10-4
Base station sectoriza-
tion (aerial view).

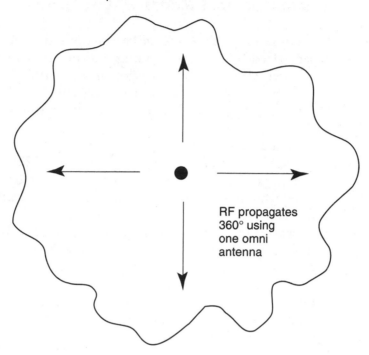

RF propagation from omni cell:
presectorization

RF propagates
360° using
one omni
antenna

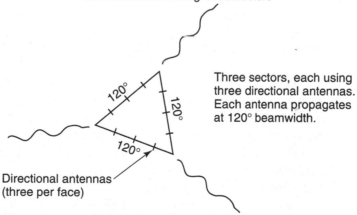

RF propagation from omni cell that's
been sectorized using three sectors

Three sectors, each using
three directional antennas.
Each antenna propagates
at 120° beamwidth.

120°

120°

120°

Directional antennas
(three per face)

10.4.4 Tower Mounting of Directional Transmit and Receive Antennas

The positioning of *directional* base station antennas is quite different in sectorized cells than it is with omnidirectional cells. When looking at a tower *platform* that has *directional* antennas mounted on it, the casual observer would notice that there are three directional antennas mounted per face on the platform. These antennas come in many shapes and sizes.

It should be noted that in some instances, for one or more sectors in a sectorized cell site, wireless carriers will install directional antennas of different gains (i.e., longer antennas) in a given sector or sectors, to achieve a higher level of gain into a specific geographic area. This is most likely done because of zoning restrictions in certain localities that may prevent the wireless carrier from installing a full-fledged base station in a certain area. To compensate for this situation, the carrier may pump up coverage into that area from the nearest cell site—from a specific sector in that cell site.

Like omnidirectional cells, directional cells use space diversity. Of the three directional antennas mounted per face in a sectorized cell, the two outer antennas are the receive antennas (RX 0 and RX 1).

Key: The length of the separation that exists between the two receive antennas is dictated by the following formula: the length between the two diversity antennas is roughly equal to 20 times the wavelength of the frequency being used.

Example: The wavelength of 1900 MHz [the U.S. Personal Communication Services (PCS) frequency band] is approximately 6 in, or half a foot. Therefore, the diversity receive antennas in a sectorized PCS cell site will be approximately 120 in apart, or about 10 ft.

Like omni cells, the diversity receive antennas receive RF signals at the cell base station from the mobile telephone. They receive the uplink signal—the mobile-to-base signal.

The antenna that is in the *middle* on any face of a sectorized cell base station is the transmit antenna. This antenna is used for the transmit channel *from the cell base station* to the mobile phone. It transmits the downlink signal—the base-to-mobile signal. See Figure 10-5.

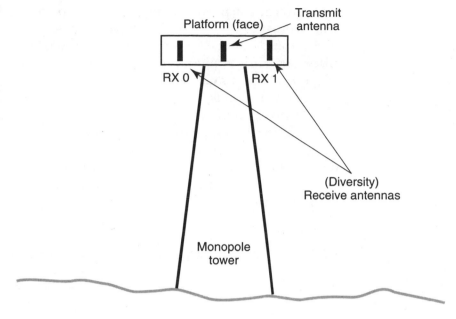

Figure 10-5
Tower mounting of
directional anten-
nas (sectorized base
station).

Platform faces on towers also vary in length depending on the height of the tower. However, each platform on any one given tower will be the same length on all sides. For example, the three platform faces on a sectorized cell site will all be the same length; but the platform faces on all the towers throughout a wireless system may be different. Theoretically, the higher the tower, the longer the faces should be; and the antennas on each face will also be farther apart. In many cases, when looking at a "duplexed" base station where there are only two antennas per face, this means that one of those antennas is used for both transmit (base-to-mobile signals) as well as receive (RX 0 or RX 1). The second antenna is used for diverse receive only. In these scenarios, it might appear as if there actually is a third antenna in the middle of the platform faces. What looks like a third antenna is actually only a *mounting post* for a third antenna.

10.4.5 Directional Gain

Using directional antennas, gain is developed by adding a *reflecting element* behind a modified omni antenna. The reflector distorts and compresses the horizontal radiation pattern at the sides, causing it to bulge forward, producing directional gain. The amount of gain developed by a directional antenna depends on the size and shape of the reflecting element and its distance away from the actual antenna.

Example: A 10-dB-gain omni antenna could be modified with a V-shaped reflector to produce a 60° horizontal radiation pattern and 17 dB of gain. This example shows that, all things being equal, a directional antenna will produce *more* gain than an omni antenna at a given location. In this example, the directional antenna produced 7 dB more gain.

Flat-plate directional antennas (panels) and parabolic dishes such as those used in microwave systems are also used to generate various degrees of horizontal beamwidth. The radiating elements placed in front of these reflectors have become very advanced technically in recent years.

10.4.6 Sectorization Impact on System Design

As with omni cell base stations, each sector in a sectorized cell can be analyzed separately for design and planning purposes. This relates to the channel sets assigned to it, the percentage of traffic it carries, century call seconds (CCS) traffic engineering information, and so on. Sectorization allows for the development of traffic studies to determine where customers actually are. It has an overall positive impact on cellular system design because it allows the carrier to *reduce* system cochannel interference. This is done by utilizing another design concept known as the *front-to-back ratio*. When sectorizing a cell, carriers usually deploy three-faced base stations, each face (sector) projecting a beamwidth of up to 120° if necessary. Today, some directional antennas on sectorized PCS cell sites have 90° beamwidths.

10.4.7 Front-to-Back Ratio in Sectorized Cells

As directional antennas are used in sectorizing omni cells, RF propagation engineers must take the front-to-back ratio into account to reduce the potential of picking up signals from cochannel sectors that are 180° to the rear. This reduces the possibility of cochannel interference.

The front-to-back ratio is defined as the ratio of the forward gain of a sector (based on the placement, or angle, of the directional antenna) to the gain 180° to the rear. Antennas with high front-to-back ratio ratings are sought by engineers and purchasing managers when trying to reduce cochannel interference in a sectorized environment. See Figure 10-6.

Figure 10-6
Front-to-back ratio in sectorized cells. RF propagation from base station A's sector 3 must not cause cochannel interference in sector 3 of base station B.

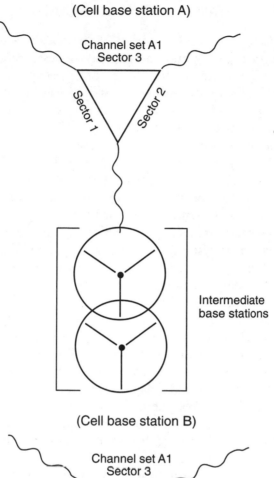

(Cell base station A)

Channel set A1
Sector 3

Sector 1

Sector 2

Intermediate
base stations

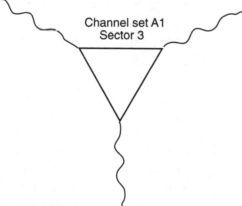

(Cell base station B)

Channel set A1
Sector 3

10.4.8 "Smart" Antenna Systems

In the late 1990s, the limitations of "broadcast" antenna technology on the quality, capacity, and coverage of wireless systems prompted an evolution in the fundamental design and role of the antenna in wireless systems. Sectorization was the first step toward increased spectral efficiency in wireless networks. The next step in this evolution was the development of the "smart" antenna.

While directional antennas and sectors multiply the use of radio channels, they do not overcome the major disadvantage of standard antenna broadcast—cochannel interference. Management of cochannel interference is the number one limiting factor in maximizing the capacity of a wireless system. To combat the effects of cochannel interference, smart antenna systems focus directionally on intended users. The goal of a smart antenna system is to increase the number of subscribers that can be serviced by a single cell at any given time. Smart antennas are designed to help wireless operators cope with variable traffic levels and the network inefficiencies they cause. These systems also allow carriers to change gain settings to expand or contract coverage in highly localized areas—all without climbing a tower or mounting another custom antenna. Wireless carriers can then tailor a cell's coverage to fit its unique traffic distribution. If need be, carriers can modify a cell's operation based on the time of day or the day of the week, or to accommodate an anticipated surge in call volume from a sporting or community event. Smart antenna systems fall into two main categories: switched beam systems or adaptive array systems. Both systems attempt to increase gain according to the location of the user. Generally speaking, each approach forms a main "lobe" (radio beam) toward individual users and attempts to reject interference or noise from outside of that main lobe.

10.4.8.1 Switched Beam Systems Switched beam systems use a finite number of fixed, predefined patterns, or combining strategies (sectors). These antenna systems detect signal strength; choose from one of several predetermined, fixed beams; and switch from one beam to another as the mobile phone moves throughout the sector. Switched beam systems communicate with users by changing between preset directional patterns, largely on the basis of the signal strength received from the mobile phone. Conversely, adaptive array systems attempt to understand the RF environment more comprehensively and transmit more selectively. In terms of radiation patterns, switched beam is an extension of the current microcellular or cellular sectorization method of "splitting" a typical cell. The switched beam approach further subdivides macrosectors into several microsectors as a means of improving range and capacity. Each microsector contains a predetermined fixed beam pattern with the greatest sensitivity located in the center of the beam and less sensitivity elsewhere. The design of such systems involves high-gain, narrow-azimuth antenna elements. The switched beam system selects one of several predetermined fixed beam patterns with the greatest output power in the remote user's channel. The system switches its beam in different directions throughout space by changing the phase differences of the signals used to feed the antenna elements or of the signals received from them. When the mobile phone user enters a particular macrosector, the switched beam system selects the microsector containing the strongest signal. Throughout the call, the system monitors signal strength and switches to other fixed microsectors as required.

The switched beam (lobe) design is the simplest smart antenna design. The main idea of this design is to measure the received signal on a finite number of predefined azimuths and choose the setting that gives the best performance.

10.4.8.2 Adaptive Array Systems Adaptive array technology represents the most advanced smart antenna approach to date. Using a variety of new signal-processing algorithms, the adaptive system takes advantage of its ability to effectively locate and track various types of signals, in order to dynamically minimize interference and maximize intended signal reception. Adaptive antenna systems approach communications between a user and base station in a different way, in effect adding a dimension of space. By adjusting to an RF environment as it changes (via the origin of signals), adaptive antenna technology can dynamically alter the signal patterns to near infinity to optimize the performance of the wireless system. Adaptive arrays utilize sophisticated signal-processing algorithms to continuously distinguish between desired signals, multipath signals, and interfering signals as well as calculate their directions of arrival. This approach continuously updates its transmit strategy based on changes in both the desired and interfering signal locations.

Among the most sophisticated utilizations of smart antenna technology is an adaptive array technology known as spatial division multiple access (SDMA). SDMA employs advanced processing techniques to basically locate and track fixed or mobile terminals, adaptively steering transmission signals toward users and away from interferers. This adaptive array technology achieves superior levels of interference suppression, making possible more efficient reuse of frequencies than the standard fixed hexagonal reuse patterns. In essence, SDMA can adapt the frequency allocations to where the most users are located. Utilizing highly sophisticated algorithms and rapid processing hardware, spatial processing takes the reuse advantages that result from interference suppression to a new level. Spatial processing dynamically creates a different sector for each user and conducts a frequency/channel allocation in an ongoing manner in real time.

Adaptive array systems provide more possibilities for frequency reuse. Unlike switched beam systems, which change the shape of the lobe (beam), adaptive array antenna systems steer the lobe as the user moves.

10.4.8.3 Summary of Smart Antenna Systems The distinctions between the two major categories of smart antennas regard the choices in transmit strategy:

- *Switched beam* antennas use a finite number of fixed, predefined patterns or combining strategies (sectors).
- *Adaptive array* antennas use an infinite number of patterns (scenario-based) that are adjusted in real time.

Although smart antenna technologies are promising many benefits, they do have some disadvantages. The main drawbacks of smart antenna systems are increased base station complexity, increased need for computational power, and more complex resource management schemes. But the single biggest disadvantage to these systems today is their cost. Currently (circa 2000), the average cost of regular base station antennas is around $500. In contrast, smart antennas cost around $4500 each. This high cost, along with the other disadvantages inherent to these antenna systems mentioned above, has precluded their widespread deployment in wireless systems today. Smart antennas have not been extensively field tested either. However, as the cost comes down, carriers will begin to deploy smart antenna systems with regularity.

With some modifications, smart antenna systems are applicable to all major wireless protocols and standards.

10.5   Test Questions

True or False?

1. _____ Each sector in sectorized cells does not have its own control channel.
2. _____ The front-to-back ratio is defined as the ratio of the forward gain of a given cell's sector to the gain at 180° to the rear.
3. _____ Cell splits are dictated when cells in a given area are consistently at 30% of capacity.

Multiple Choice

4. The sectorization process:
   (a) Subdivides an omnidirectional cell into three separate subcells using directional antennas
   (b) Minimizes cochannel interference
   (c) Optimizes the frequency-reuse plan
   (d) From a graphical perspective, converts a regular circle into a three-sectioned pie chart
   (e) All of the above
   (f) a and d only

5. Directed retry:
   (a) Defines the process of creating a smaller cell inside a larger cell by manipulating the power levels of certain radio channels
   (b) Defines the process of routing call origination attempts to nearby cells
   (c) Defines the process of subdividing an omni cell into three equal subcells
   (d) Defines the process of creating a new cell by splitting the distance between the center of two existing (or planned) cells into two or more cells

6. What technique describes the process of (figuratively) building a small cell *inside* a larger cell by manipulating RF power levels of alternate channels at neighboring cell sites?
   (a) Quad-level sectorization
   (b) Franchising
   (c) Overlay/underlay
   (d) Directed retry

Cellular Regulatory
Processes

The FCC is the regulatory body for the cellular and PCS industries. The FCC issues market licenses to cellular and PCS carriers. Also, cellular radio equipment must be "type-accepted" by the FCC prior to shipment by manufacturers to wireless carriers.

11.1 Construction Permits

During the first 12 years of the cellular industry's existence, the FCC issued regulations that had to be adhered to when cellular carriers obtained licenses to build out their markets. Some of these regulations have now gone by the wayside because cellular markets have already been built out by carriers nationwide. A brief review of these regulations is given below. There were several levels of construction permits (CPs) issued to cellular carriers, once they were licensed, in markets that had not yet been developed.

11.1.1 18-Month Fill CP

The FCC required that, if a cellular carrier acquired a license for a cellular market that had never had cellular service before, at least one cell site had to be activated within 18 months of the operating license being granted to that cellular carrier. The purpose of the 18-month CP was the FCC's way of determining if a carrier could build a base station and make it operational.

11.1.2 5-Year Fill CP

The FCC required that, at the end of 5 years from the grant of the cellular license, the cellular carrier had to cover that territory within its market that it deemed "profitable." What this means is that cellular coverage had to be provided to those portions of the market where there was enough population to warrant construction of cell sites. Any geographic territory remaining uncovered by the FCC-required 32-dBu contour after 5 years was subject to license applications by other cellular carriers.

11.1.3   10-Year Renewal

An FCC cellular license is issued for a period of 10 years. At the end of this period, a cellular operator must make a statement as to its qualifications to continue to be the cellular licensee in a given market. The cellular carrier should have a clean record in dealing with the FCC on regulatory issues and with the public on providing "quality service." Complaints against cellular carriers can be filed and reviewed, but licenses are usually reissued. (The current rate of reissuing licenses is 100%.) There are also methods for applying to serve unserved territory, known as Phase One and Phase Two Applications.

11.2   FCC Forms

The FCC requires that certain forms be filed with it at certain points in the process of building cell sites and towers. As of the late 1990s, FCC forms can be filed electronically through a licensing web site managed by the FCC. Initially, from 1983 until around 1995, cellular carriers had to obtain construction permits from the FCC before *any* cell sites in their markets could be built. But today, for an internal cell (a cell within a carrier's market that is not situated on a border with an adjacent cellular market), no forms are required to build new cell sites.

11.2.1   The 600 Form

This form is used for applying for "Minor and Major Modifications" to a carrier's cellular market. It is used to notify the FCC that a border cell has been activated. It is required because the border cells add territory to the carrier's market. Once this form has been sent to the FCC, a reply is received within 90 days.

11.2.2   The 489 Form

This form is labeled "Notification of Status of Facilities." It is used to notify the FCC that a cell site has been built. It is sent to the FCC to complement the filing of the 600 form. When the activity that was detailed in a 600 form has been completed, a 489 form must be filed with the commission.

11.2.3 The 494 Form

This form must be filed with the FCC for microwave radio licensing. Once filed, it takes about 1 to 2 months to receive a reply from the FCC. Frequency coordination is also required to install microwave radio systems, and this coordination must be taken into account along with the filing of the 494 form in determining the total time needed to activate a microwave radio system.

11.3 FAA Forms

11.3.1 The 7460-1 Form

The FAA 7460-1 form must be filed to show tower site locations. The form must show the base elevation of the site above mean sea level (AMSL) and the overall height of the tower and antennas so the FAA can determine if the tower is a hazard to air navigation. Wireless carriers must also submit a 7.5-minute USGS map showing the cell site with 7460-1 forms delivered to the FAA. The FAA will study the application and notify the cellular carrier if the tower must be lighted and/or marked. It takes the FAA about 1 month to process a 7460-1 form. Once the FAA receives the 7460-1 form, it will also notify a carrier if a 7460-2 form must be filed.

11.3.2 The 7460-2 Form

If a tower exceeds certain heights in certain locations, the FAA 7460-2 form is used to inform the FAA when a carrier will begin and end construction of the tower, so the FAA can notify area airports.

11.4 The 32-dBu "Carey" Contour

The 32-dBu contour is a formula used to depict a contour that serves as a legal definition of a cellular carrier's market. The formula takes several factors into account: tower/antenna height, ERP, and the average terrain height AMSL. The 32-dBu contour was developed by an FCC engineer, Roger Carey, to develop coverage areas of commercial radio stations, paging sites, and specialized mobile radio (SMR) sites.

The purpose of the 32-dBu contour is regulatory in nature. It ensures fairness among and between cellular carriers in that it makes all carriers compute coverage in the same way. The 32-dBu contour defines *regulatory* cell site coverage, but not necessarily *real-world* coverage. It is a poor approximation of actual coverage because it ignores ground clutter and urban sprawl. In hilly terrain, the 32-dBu contour overprojects coverage, because terrain features aren't factored in. In flat areas, it underprojects coverage substantially.

Cellular carriers used to have to show the 32-dBu contour for every cell in their system, but it is now only required for boundary cells.

11.5   Extension Agreements

Extension agreements are agreements between two cellular carriers which denote how much RF signal coverage can propagate across market boundaries between the two carriers. These agreements become necessary when a cellular service provider's 32-dBu contour crosses into an adjacent cellular market. In order for that contour to remain inside the adjacent carrier's market, a formal signed agreement must be obtained. Extension agreements are legally binding documents.

There may not even be any real-world coverage on the border between two cellular markets because of terrain factors, but extension agreements are necessary because the 32-dBu contour is a *legal* definition of a cellular carrier's service coverage area.

Extension agreements are important because they regulate the points at a carrier's market border where call origination attempts are carried on a subscriber's home system instead of being carried on the adjacent market carrier's system. If a call attempt is picked up by the neighboring cellular system, the call is then considered a roaming call, and the rates charged to the subscriber can be substantially higher.

Key: Circa 1998, extension agreements have become rare among wireless carriers because of a relaxation of FCC regulations.

11.6   System Information Updates

System information updates (SIUs) define, for the FCC, geographical areas that are unserved by cellular carriers. The FCC requires maps showing the 32-dBu contour for a carrier's contiguous market serving area 60 days before the 5-year CP expires and at the actual CP expiration date.

11.7   Test Questions

True or False?

1. _____ The 18-month construction permit (CP) requires that, at the end of 18 months, a cellular carrier must have provided coverage to that territory within a new cellular market that they deem "profitable."

Multiple Choice

2. Cellular (market) licenses are issued by the FCC for a period of how long?
   (a) 5 years
   (b) 2 years
   (c) 18 months
   (d) 10 years

3. Which form must be filed with the FCC to notify it that a "border" cell has been activated between two different cellular carriers' adjacent markets?
   (a) The 494 form
   (b) The 600 form
   (c) The 489 form
   (d) The 7460-2 form

4. What represents the legal definition of a cellular carrier's service coverage area?
   (a) The real-world contour
   (b) The extension agreement
   (c) The 32-dBu contour
   (d) The FCC 489 form
   (e) None of the above

5. Which form must be filed with the FCC in order to activate a microwave radio system?
   (a) Form 552
   (b) Form 600
   (c) Form 489
   (d) Form 494
   (e) 7460-1 form
   (f) None of the above

CHAPTER 12

Enhancers and Microcells

12.1 Overview

Enhancers, sometimes known as *repeaters,* are special radios that are used to boost cellular RF signals in small areas where RF coverage is inadequate (e.g., tunnels, malls, dead spots, and areas between cell sites). Enhancers do *not* contribute any capacity (in erlangs) to a wireless system. Due to advances in technology today, enhancers can be placed in small cabinets and mounted almost anywhere. Buildings to house equipment are not necessary; the main components are radio equipment and an antenna. Enhancers use RF channels from donor cells to increase the coverage area that's capable of being processed by the donor cell.

There are two main types of enhancers: *broadband* enhancers (also known as "pass-through" enhancers) and *channelized* enhancers. Channelized enhancers *can be* "translating" enhancers. Translating enhancers are rarely used today—they're not used as much as they should or could be used—to fill in coverage "holes" because they have limited functionality.

Key: The geographic location of an enhancer (but not the equipment) is capable of being converted to an actual cell base station.

12.2 Functional Types of Enhancers

The two predominant types of enhancers are described below.

12.2.1 Pass-through Enhancers

Pass-through enhancers are also known as *broadband enhancers.* These enhancers use channels that are fed from an omnidirectional donor cell. A highly directional antenna known as a *pickup antenna* is used at the enhancer location to receive the signals propagated from the omni antenna at the donor cell. The highly directional antenna ensures that the pass-through enhancer's pickup antenna receives *only* the assigned channels from the donor cell. Pass-through enhancers can enhance channels from either the A carrier spectrum or the B carrier spectrum. Therefore, the use of directional antennas at the enhancer location is

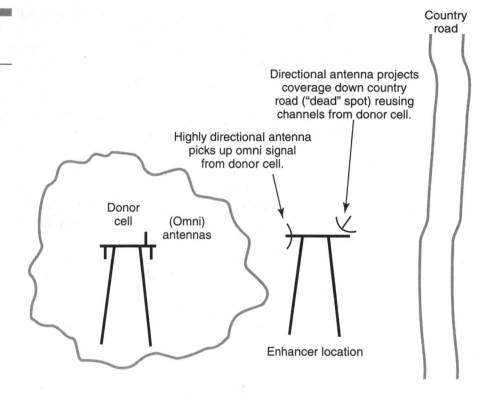

mandatory according to the FCC. Subscriber traffic that is picked up and carried within the enhancer's coverage area—via call originations or call handoffs—is also routed to the donor cell through these directional antennas for call processing. Directional antennas are then also used at the pass-through enhancer location to provide RF coverage to very specific geographic areas around the enhancer's location, where "dead" spots may exist (e.g., down a highway corridor in one direction). If desired, a wireless carrier could place *more than one* directional antenna at an *enhancer location* to provide coverage in several directions (e.g., up *and* down a highway corridor).

The power used at a pass-through enhancer location could be high or low, depending on the type of amplifier used. See Figure 12-1.

12.2.2   Translating Enhancers

Translating enhancers are enhancers that accept an RF feed from an omnidirectional donor cell and use a different channel set than the channel set at the donor cell. RF from a donor cell is received at the translating enhancer through a directional antenna. At the enhancer location, an omnidirectional antenna can be used to project RF into the desired coverage "hole" to be accessed by wireless subscribers. The enhancer itself translates a given RF channel it receives from the donor cell into a different RF channel, and propagates that new channel into its designated coverage area. For example, the enhancer will receive channel 400 from the donor cell, translate that channel to channel 500, and project it into the coverage area. This scenario begs the question: If you are using an entirely different channel set at the enhancer location, why not just install a whole new cell site at the enhancer location? The reason is that installing an enhancer rather than a full-fledged cell site saves hundreds of thousands of dollars. All the carrier installs at the enhancer location is a tower and an enhancer.

A different channel set is used at the enhancer location because it reduces the need for a high degree of isolation (signal separation) between the "link" antenna that receives signals from the donor cell and the antenna that broadcasts to the subscriber. A different channel set is also used in order to avoid cochannel interference. See Figure 12-2.

12.3 Microcells

Wireless carriers are struggling with an unusual paradox today, circa 2000. Fickle consumers demand consistent, dependable wireless services, yet they do not want to make the sacrifices necessary to accommodate traditional base station installations. As mentioned in Chapter 4, "The Cell Base Station," this is the NIMBY (not in my backyard) phenomenon at work. In 1984, when the wireless industry was emerging as a commercial business, the size and location of base stations were not issues. By 1986, there were only 1531 cell sites in the United States, according to the Cellular Telecom Industry Association (CTIA). By 1998, that number had increased to over 67,000 cell sites. Because of the NIMBY factor, municipal governments began passing zoning ordinances as a way to block base station construction, particularly in residential areas.

The Telecom Act of 1996 allowed the federal government to thwart many of the tactics used to dissuade wireless carriers from network expansion. However, the act still gives final authority on the placement, construction, and modification of wireless services facilities to state and local governments. This motivated wireless carriers to look for base stations that

Figure 12-2
Translating
enhancers.

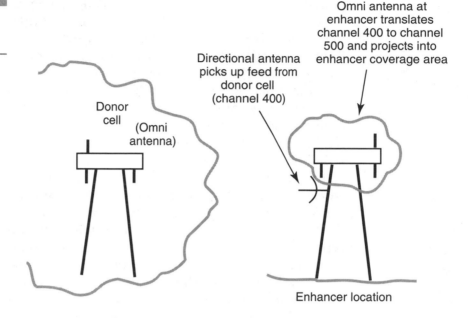

Donor
cell
(Omni
antenna)

Directional antenna
picks up feed from
donor cell
(channel 400)

Omni antenna at
enhancer translates
channel 400 to channel
500 and projects into
enhancer coverage area

Enhancer location

imposed smaller footprints on their surroundings. Like the trend
toward smaller handsets, the industry is seeing a trend toward small
wireless infrastructure equipment. Microcell technology is fostering this
trend with new technology patents that provide more features, greater
range, and expanded capacity to carriers. Once seen as a niche product
to fill gaps in coverage areas and as a replacement for enhancers, modern
microcell technology gives carriers a way to deploy base stations quickly
and cost-effectively, while minimizing impact on the surrounding area.
These small microcellular base stations can provide the same capacity
and coverage area as that of their traditional-size counterparts, "macro-
cellular" base stations. Carriers can plug-and-play microcells to fit their
design needs, while significantly reducing initial deployment and recurring network costs.

Physical size and weight are the primary distinguishing features
between microcells and traditional base stations. Traditional base stations typically weigh close to 1000 pounds (half a ton) and occupy 54
cubic feet. They require weatherproofed, air-conditioned enclosures and
environmental aesthetics. These large, fixed structures are difficult to
disguise, particularly in residential areas where they are usually least
welcome.

The new generation of microcell base stations is ruggedized boxes
that generally measure about 6 cubic feet and weigh less than 100

pounds! Microcell base stations are designed to withstand temperatures ranging from −40 to 158°F. They require no additional heating or cooling systems like a traditional base station would. Because of the small form factor, microcells are readily installed wherever they can be easily and imaginatively concealed: on the side of a building, in a tunnel, on top of a utility pole, or under a bridge. A microcell base station's unobtrusive size is ideal for suburban residential areas where zoning restrictions tend to be the most stringent. Outside of residential areas, microcell base stations can be used to improve or extend coverage in office buildings, enclosed shopping malls, sporting areas, airport terminals, and other indoor locations where people are likely to engage in heavy wireless phone use. Because no buildings or special power sources are necessary, microcell installation time is reduced from days to hours.

The range of a microcell is 2 mi or less, depending on the antenna type used. Outdoors, directional antennas are usually used for microcells. Indoors, omni antennas, directional antennas, or "leaky" coax cable may be used for microcell coverage. Leaky coax is a coaxial cable that has holes cut into its shielding at regular intervals to allow the RF signal to propagate outward. Leaky coax is very expensive compared to antennas.

There are now special low-power cells, called *picocells,* which are extremely low-powered. These cell sites are mostly used in wireless in-building systems [e.g., wireless private branch exchanges (PBXs)], very congested downtown pedestrian areas, or even shopping malls.

12.4 Cost Comparisons

Wireless carriers can always weigh several options when deciding what type of radio equipment they deploy in any given area. One of these options is price. The cost of installing a microcell is about one-fifth the cost of installing a large, traditional base station.

In summary, at present (circa 2001) microcells cost between $60,000 and $100,000 to install. The average traditional base station can cost between $125,000 and $750,000. The average enhancer costs between $15,000 and $50,000.

12.5   Test Questions

True or False?

1. _____ Picocells are mostly used in wireless in-building systems.

2. _____ Pass-through enhancers are also known as *broadband* enhancers.

3. _____ Both pass-through and translating enhancers use channels from "donor" cells.

Multiple Choice

4. Translating enhancers can use what type of antenna to propagate coverage into the enhancer's coverage area?
   (a) Omnidirectional
   (b) Bidirectional
   (c) Directional
   (d) Downcasting
   (e) Both a and c
   (f) None of the above

5. Microcells:
   (a) Have all the functionality of a regular cell site, but with lower power output
   (b) Can sometimes be used to cover large geographic areas
   (c) Are used to fill small coverage holes such as buildings, shopping malls, and convention centers
   (d) Have a range of 10 mi or greater
   (e) a and c only
   (f) All of the above

Design Tools and Testing Methods

13.1 Propagation Modeling Tools

Today's software tools for predicting RF propagation use highly complex mathematical formulas developed over years of field testing. These tools have been developed by using the formulas to *try* to predict what signal levels will appear to be in any given terrain situation.

The output of computer propagation models is usually a predictive graphical plot of cell site propagation that can be merged with a map software program to show predicted RF coverage in an area, along with the local roads and highways, all in one output. The real world won't fit into any propagation models because the real world is *constantly chang-ing:* Trees are planted and torn down and lose their leaves for 5 months of the year, buildings are torn down or erected, trucks drive by, highways are built, etc. All these things—anything that sticks up above the ground—will affect wireless RF coverage. No software model could possibly keep up with changes in terrain features or predict RF propagation to 100% accuracy in a constantly changing world. There is simply insufficient time to implement this information into propagation models. Predictive modeling software programs can only have a given percentage of accuracy, depending on how many factors of terrain and clutter are taken into account when the programs are executed. See Figure 13-1 for an example of a real-world predictive plot from a computer program that simulates RF coverage over specific geographic areas in Indiana.

13.2 Drive Tests

Drive tests describe the process of a technician driving through a wireless market to check the RF signal levels as the vehicle progresses from one cell to the next, using signal measuring test equipment. RF signal measurements that determine when call handoffs are dictated are also recorded and turned in to RF engineers for analysis. Even drive tests are actually limited in their effectiveness. Drive tests are *more* accurate than a propagation model, but still not the complete answer for predicting how RF signals will propagate at any given time.

Key: Drive tests give an engineer only a snapshot in time of how RF is propagating, from a specific point.

Figure 13-1
Predictive plot using RF propagation software modeling tool. The lighter gray areas indicate progressively fading RF coverage. Car-mounted mobiles will always obtain higher degrees of coverage than hand-held portable phones.

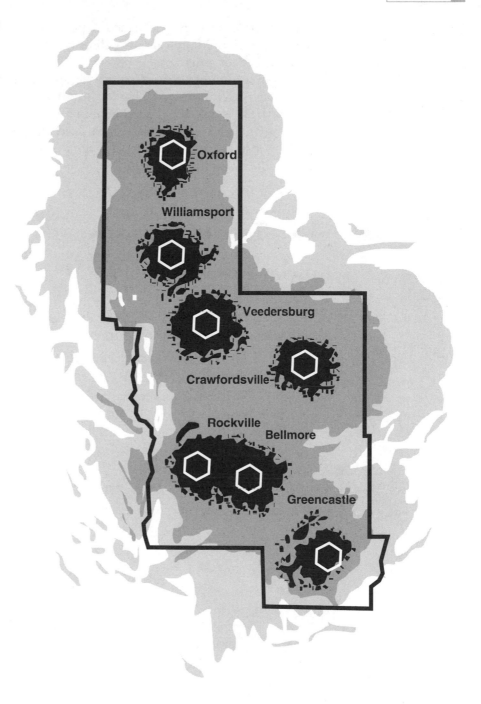

How often drive tests are performed depends on the ownership of the market in question. Many carriers will only do drive tests when they do a major "retune" of a market.

It is a physical trait of RF that, at one spot, signal levels could vary greatly at any given moment. That is also a natural characteristic of signals in the UHF band. After powering up a portable wireless phone that comes with an LCD screen and a signal level indicator, move the phone around or just keep looking at the signal level indicator. The signal strength will constantly fluctuate.

13.3   Test Questions

True or False?

1. _____ Drive tests give a cellular engineer only a snapshot in time of how RF is propagating from a specific point.

2. _____ The real world fits easily into propagation models because the real world rarely changes.

3. _____ Drive tests are more accurate than a propagation model.

14

The Mobile Switching Center

14.1   Overview

The *mobile switching center* (MSC) is the location of the switch that serves a wireless system. It is similar in function to a class 5 end-office switch in the public switched telephone network (PSTN). The MSC is also known as the mobile telephone switching office (MTSO), a term used largely by cellular carriers. The term MSC became popular among the PCS carriers, who started building their networks in the late 1990s. All wireless system base stations are connected electronically to the MSC.

The primary purpose of the MSC is to provide a voice path connection between a mobile telephone and a landline telephone, or between two mobile telephones. The MSC serves as the nerve center of any wireless system. Functionally, the MSC is supposed to appear as a seamless extension of the PSTN from the customer's perspective. The MSC is composed of a number of computer elements that control switching functions, call processing, channel assignments, data interfaces, and user databases. Wireless switches are some of the most sophisticated switches in the world today.

Predominant makers of MSCs include Nortel, Ericsson, Nokia, Motorola, DSC, and Lucent Technologies.

14.2   MSC Functions

General, high-level functions of the MSC include the following:

- *Tracking.* The MSC tracks all wireless users in its system through a process known as *autonomous mobile registration,* and constantly monitors the technical health of all cell base stations connected to it. If there is an equipment problem within a cell, the MSC will be notified through an alarm process. The MSC will then forward the alarm information directly to the network operations center (NOC) for processing and action. This will usually take place over a carrier's own wide-area network (WAN).

- *Call processing.* The MSC handles processing of all wireless system calls. This includes mobile-originated calls, where the MSC coordinates the seizure of trunks to the PSTN. It also includes mobile-terminated calls, where the MSC coordinates and activates the paging function to locate the mobile phone (see Chapter 19,

"Cellular Call Processing"). For mobile-terminated calls, the MSC must initiate paging in areas where the mobile station's signal has been recorded in a database via the mobile registration process.

- *Paging function.* The paging function is the means by which the MSC locates mobile subscribers for purposes of connecting land-to-mobile or mobile-to-mobile telephone calls. For example, the MSC will try to locate the mobile by initiating a paging signal in the last 10 cell sites that the wireless user's signal was registered. If the user is not located in those 10 cells, the MSC may initiate a *regional* page to locate the mobile (i.e., 50 cell sites). If the user has still not been located, the MSC may initiate a *global* page, where it sends a paging signal to all base stations in its system (i.e., 200 to 300 cell sites). This process is very carrier-specific, and would take only milliseconds to occur.

- *Call handoff.* Call handoff is the process where the MSC coordinates the transfer of a call in progress from one cell to another cell. The call handoff function is tightly controlled by the MSC.

- *Billing.* Customer billing is controlled by the MSC. Through what are known as *automatic message accounting* (AMA) records, the switch records call detail data onto either a magnetic disk or tape drive.

- *Roamer data.* Through MSC databases that are known as *home location registers* (HLRs) and *visitor location registers* (VLRs), the switch can keep track of subscribers who are roaming from another wireless system (see Chapter 20, "Roaming and Intercarrier Networking"). Once the MSC has determined that a roamer is in its system, it will know to direct the billing information to a special data storage file, where ultimately a roaming clearinghouse will properly distribute the monies to other wireless carriers.

- *Traffic and call processing statistics.* The MSC collects data on all call processing functions, to include traffic statistics on cell radio trunk groups, trunk groups that connect the wireless system to the PSTN, call handoff statistics, dropped calls, and maintenance statistics (cell and MSC alarms). See Figure 14-1.

MSCs in some wireless systems will have *base station controllers* (BSCs) in place, which essentially serve as front-end processors to the switches themselves. The purpose of the BSC is to off-load some of the base station and call management functions from the switch, so the switch can perform more switching-related and database-related functions. This

Figure 14-1
MSC functions.

To cell sites   **To PSTN**

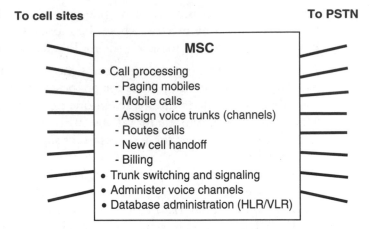

MSC architecture—use of base station controllers—will depend on several factors. First, some wireless switch manufacturers, such as Nortel and Lucent, include BSCs as part of their standard functional infrastructure. Second, GSM standards dictate the use of a BSC as part of the standardized design and operation of GSM digital wireless systems. The placement and use of base station controllers is part of the GSM standard.

At some point, there are many instances where wireless carriers must investigate the need to relocate MSCs to other locations that would be more sensible. This analysis is usually required as a given market's network grows rapidly over a short period of time. The requirement to relocate, or move, an MSC could result from any one of the following factors:

1. Another entirely new MSC may be required to serve a portion of a given market, given a substantial increase in the number of cell sites that have been constructed over a period of time. A single MSC may not be able to handle the traffic load when many new cell sites are deployed in a wireless market. If a new MSC is being installed, that activity may demand that the wireless carrier relocate the existing MSC to an area that will make the traffic load in the market more balanced, in terms of existing cell base station locations and the existing PSTN interconnection architecture in a certain market.

2. If there are key interconnections to the PSTN into a given central office (i.e., an access tandem), the wireless carrier may want to move the MSC so that it is physically closer to the access tandem. This will decrease the monthly recurring cost for facilities (circuits for fixed network and interconnection). It will also decrease the odds that a substantial fiber outage will take place, effectively bringing the MSC and the entire system to its knees.

3. If it is difficult to obtain interconnection or fixed network facilities at an MSC location, the wireless carrier may wish to move the MSC closer to copper or fiber facilities' rights-of-way, or closer to a telephone company central office itself.

14.3   Test Questions

1. List and briefly explain the seven key functions of a wireless carrier's switch.

Multiple Choice

2. The requirement to add, relocate, or move an MSC does *not* result from which one of the following factors?
    (a) Another entirely new MSC may be required to serve a given market because of a substantial increase in the number of cell sites that have been constructed over a period of time.
    (b) From an interconnection perspective, it is desirable to move the MSC so that it is physically closer to the access tandem to decrease the monthly recurring cost for facilities.
    (c) Because the in-market competition has added or moved an MSC closer to a specific interconnection point.
    (d) It is difficult to obtain interconnection or fixed network facilities at an MSC location, and the wireless carrier therefore wishes to move the MSC closer to copper or fiber facilities' rights-of-way, or closer to a telco central office itself.

The N-AMPS Standard

15.1 Overview

Narrowband AMPS (N-AMPS) is the name given to a cellular air interface specification developed by Motorola. N-AMPS is a direct descendant of Narrowband Total Access Communication System (N-TACS), which is a cellular air interface standard that was enormously popular in Japan into the early 1990s. N-AMPS systems are only installed in systems that use Motorola equipment, exclusively.

360 Degrees Communication (now merged with AllTel Mobile) rolled out N-AMPS in its Las Vegas market in 1992, and now has N-AMPS systems operating in Raleigh, North Carolina, and Tallahassee, Florida.

15.2 The N-AMPS Configuration

Although N-AMPS is *not* truly a digital technology, it is often classified with the digital services since it employs a *digital signaling channel.* Use of this channel allows for additional security and features, as described below.

The N-AMPS design uses each 30-kHz AMPS cellular channel and splits it into three 10-kHz subchannels. This configuration is made possible by the digital signaling channel inherent in the N-AMPS configuration. Each cellular transmit or receive channel in N-AMPS is allocated a 10-kHz channel in N-AMPS, just as a cellular call is allocated a 30-kHz channel in the AMPS system.

15.2.1 Benefits of the N-AMPS Configuration

The benefits afforded by implementation of an N-AMPS cellular system are as follows:

- *Greater capacity.* The N-AMPS three-for-one channel split results in a threefold increase in system capacity without the overhead of cell splitting and all its attendant headaches (e.g., frequency coordination).
- *Compatibility.* N-AMPS is compatible with the AMPS system in that the 30-kHz control channel can still be used and dual-mode mobile telephones can be built to operate under both standards.

The N-AMPS standard actually specifies operation over both AMPS and N-AMPS channels. Thus, an N-AMPS-compatible mobile telephone may be directed to an AMPS channel, depending on the resources available (e.g., no idle N-AMPS channel available) at a dual-mode cell site, and vice versa.

■ *Automatic repeat request (ARQ) signaling.* The 200-bps signaling afforded by N-AMPS systems does away with the 10-kHz tones that AMPS uses for SAT tone. (See Chapter 19, "Cellular Call Processing.") It is now possible for the mobile phone to acknowledge orders it has received from the cell base station. This is especially important in the area of call handoff, where the AMPS handoff order confirmation of 50 ms of 10-kHz signaling tone is often missed because of interference or incorrect calibration of the cell base station tone detectors.

■ *Improved call control.* The N-AMPS standard provides for a feature known as *mobile reported interference* (MRI). The cell base station can request that the mobile telephone send in a measure of the forward voice channel [traffic channel (TCH)] signal strength, as well as a measure of the number of errors in the 200-bps signaling stream. The cell base station may then use this information as further input to the call handoff and power control detection software.

■ *Applicability to portable service.* N-AMPS requires no significant change from AMPS in terms of technology or packaging. Small, cheap, and lightweight N-AMPS portable handsets were available from the first day of service.

■ *Preloading.* Since N-AMPS mobile telephones are dual-mode, subscriber equipment to support N-AMPS may be sold to the general public well in advance of N-AMPS service actually becoming available. This way, N-AMPS service providers are assured of high utilization of N-AMPS base station equipment when they decide to deploy it.

15.2.2   Disadvantages of the N-AMPS Configuration

The very nature of N-AMPS degrades the signal quality because only one-third the normal AMPS channel bandwidth is used.

15.3   N-AMPS Digital Features

N-AMPS also incorporates several digital features that AMPS does not have, including the following:

1. *Short message service.* Alphanumeric messages of 14 characters or less may be sent on the forward channel in a point-to-point or multipoint mode, thus combining paging functions with cellular service.

2. *Enhanced paging.* N-AMPS gives carriers the ability to do two-way paging.

3. *Voice mail notification.* N-AMPS has the capability to activate a message waiting light on users' cellular telephones to notify them that they have voice mail messages.

15.4   Test Questions

True or False?

1. _____ Although N-AMPS is not truly a digital technology, it is often classified with the digital services since it employs a digital control channel.

2. _____ The N-AMPS standard does not specify operation over both AMPS and N-AMPS channels

Multiple Choice

3. The N-AMPS standard:
   (a) Subdivides a 30-kHz AMPS channel by splitting it into three 10-kHz subchannels
   (b) Is compatible with the AMPS standard
   (c) Incorporates digital features that the AMPS standard does not
   (d) Has been deployed nationwide by all cellular carriers, and in Canada
   (e) All of the above

(f) a, b, and c only

(g) a and b only

4. Which of the items listed below is not an enhanced feature offered by the N-AMPS standard?

(a) Enhanced paging

(b) Voice mail notification

(c) Multimedia capabilities

(d) Short message service

5. In an N-AMPS system, it is possible for the mobile phone to do what with orders it has received from the cell base station?

(a) Preempt

(b) Circumvent

(c) Relay

(d) Acknowledge

The Fixed Network and System Connectivity

16.1   Overview

All cells in a wireless system must be electronically connected to the MSC. This is why a fixed network *overlay* is required for all wireless systems. Because of call handoff, the wireless system also requires that all cells must be able to communicate with all other cells, electronically (logically) through the MSC. This is how the MSC coordinates signal strength measurements that are used in the call handoff process—to determine when handoffs are required.

All cells must always be able to signal all other cells, and this is another reason why the hex metaphor was chosen as the ideal tool by which to design the wireless system. The hexagon design accounts for conceptual and physical overlapping between all cells because wireless coverage is depicted by circles from an engineering standpoint.

The main objectives of the fixed network are to satisfy capacity demands and provide reliable service. This can be achieved if the following criteria are incorporated in the network's design:

- All routes must be sized to meet service demands.
- Traffic must be routed in the most economical manner.
- Survivability must be built into the network.

It is critical that a transmission plan is established indicating current and potential future cell sites.

Key: The fixed network can consist of connections based on leased facilities such as copper cable or fiber optics. The fixed network may also consist of microwave radio links, or a combination of any of these options.

However, the use of satellite links in the fixed network is rare because of their high cost. All fixed network connectivity, regardless of the actual medium that is employed (e.g., microwave or leased line), is accomplished via DS1 (T1), DS3, or OC-*n* transmission systems.

Key: Each wireless call occupies one DS0 of bandwidth over the fixed network. The process of transporting traffic across a wireless network back to the MSC for switching is known as *backhauling.*

16.2   Network Configuration Options

There are several methods by which cell base stations can be connected to the MSC, similar to the configuration options that exist for wide-area networks (WANs). These options are a star formation, a ring formation, or a daisy-chain formation. The fixed network methodology that is implemented may depend on the nature of the wireless market (existing infrastructure) or the policies of the wireless carrier.

Key: All interconnections to the PSTN must also ultimately connect into the MSC. These interconnections can traverse the *same* fixed network that the cell base stations use to connect to the MSC.

16.2.1   Star Configuration

Network topologies that use only star formations are rare today, in any type of network, because they do not allow for (nodal) transport diversity. When star topologies *are* used, it is usually in microwave networks, where the star portion is subtended by a hubbed portion that contains path diversity between the major hub nodes themselves. In effect, this topology is a combined star-ring design (see Chapter 17, "Microwave Radio Systems"). Star formations that connect all cells to the MSC are almost unheard of, because of the potentially high cost of *independently* connecting every single cell base station to the MSC. This is especially true for rural cellular markets, where the geographic expanse of the market is much larger than that of an MSA. Economically, it would not make sense to have in place many long, stand-alone network links from each cell site to the MSC in a rural market. It might make more sense to install another MSC in the market. See Figure 16-1.

16.2.2   Ring Configuration

When referring to ring configurations in the wireless fixed network, the rings can be implemented in one of two ways: using Synchronous Optical Network (SONET) rings via leased facilities; or via microwave radio

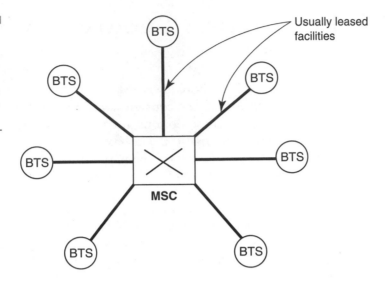

networks, which are usually designed, built, owned, and maintained by
the wireless carriers themselves.

When wireless carriers choose the SONET option for the fixed network, they will lease the facilities from either the incumbent local
exchange carrier (ILEC) or a competitive local exchange carrier
(CLEC). These systems will have transmission capacities of OC48, and
increasingly OC192. SONET systems are deployed in ring formations
to achieve diversity through what is known as *path switching*. This
means that the system always has two diversely routed transmission
paths available to it, and it is constantly checking these paths to
ensure that traffic is routed over the path with the best overall signal.
If (voice) traffic is routed over path A, and path A's signal degrades,
then traffic will be routed in the reverse direction over path B—it
will be "path switched." Obviously, the most glaring example of signal
degradation on any given path is a complete fiber cut, also known in
the industry as "backhoe fade" because the use of backhoes is what
usually accounts for major fiber-optic cable cuts. This redundant
capability of SONET rings is also known as fault tolerance, or "survivability." It's important to note that the SONET ring option is widely
available in urban areas, so it is therefore best suited as an option in
an urban cellular market (MSA). It is not widely available in most
rural areas because the demand (potential network traffic) is not
there, which makes the economic case for telcos and CLECs to build
SONET systems in a rural area a negative revenue proposition. See
Figure 16-2.

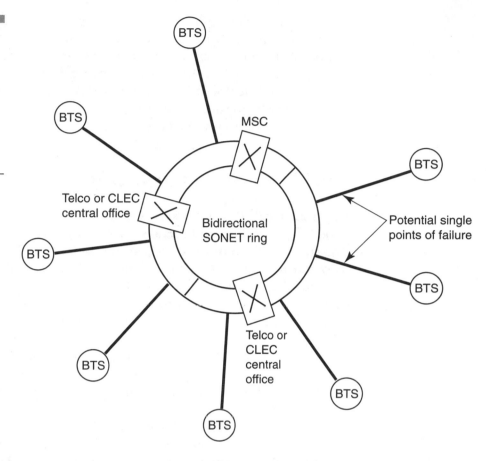

Figure 16-2
Fixed network: ring configuration (SONET). Each switching center is a node on the ring, including the mobile switching center (MSC). BTS = base transceiver station.

Key: When wireless carriers do employ the SONET ring option, the MSC itself is implemented as a "node" on the ring, as are multiple ILEC or CLEC central offices and/or wire centers.

When wireless carriers use SONET rings in their fixed networks, it's important to note that although the ring itself is fault tolerant to failure ("survivable"), the links from each base station to the entry point of the ring (telco or CLEC central offices) are in most cases not survivable. This is what is known as the "single point of failure" in a (fixed) network. Yet, this still means that in the event of a major failure of leased facilities on the ring itself or in any base station's connection into the ring, that only one base station connection would be lost instead of many base station connections. See Figure 16-2 for a graphical depiction of the ring architecture in a wireless fixed network.

The microwave radio option, when employed, will usually be used in a rural environment due to easier zoning laws and general ease of implementation. For example, line of sight is usually easier to obtain in a rural area versus an urban area. Microwave radio systems can also be designed and deployed in ring configurations, very similar to SONET systems that are leased from the local phone company or a CLEC. (See Chapter 17, "Microwave Radio Systems.")

16.2.3   Daisy-Chain Configuration

In larger rural cellular markets (RSAs), a daisy-chain architecture may be a good deployment option because of the large expanse of land that must be traversed for cells to connect to the MSC. In a daisy-chain fixed network architecture, all cells throughout a wireless market connect *to each other* as they home in toward the MSC (Figure 16-3). With this configuration, the cumulative amount of trunks on any link between two cells gradually becomes greater and greater as the links move toward the MSC. This is due to the incremental effect of adding more and more cells and trunks to the network as it approaches the MSC. For instance, the link with the largest capacity (and therefore the largest overall amount of trunks) will be the last link in the chain between the final cell in the chain and the MSC itself.

Note: A stand-alone link that runs off the main backbone of a daisy-chain network is known as a *spur.*

It is very important to plan for growth when the daisy-chain method is used for the fixed network. If planning is not done intelligently, the capacity on any one of the cell-to-cell links could be maximized as base stations are added and the wireless network grows, depending on where *new* cell sites are placed. This situation could require a major reconstruction of entire sections of the daisy chain, or even the whole chain itself. Or the wireless carrier could add another strategically placed MSC at a point toward the distant (opposite) end of the main backbone of the daisy chain.

One predominantly rural, nationwide cellular carrier uses microwave radio links, laid out in a daisy-chain fashion, to connect all of its cells to the MSC. In this scenario, the daisy-chain architecture has been very effective and optimized the cellular carrier's network infrastructure.

Figure 16-3
Fixed network: Daisy-chain configuration.

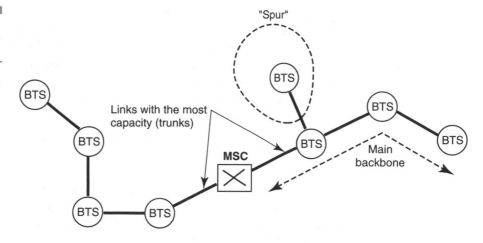

Key: The daisy-chain and ring configurations also allow a cellular carrier to attain economies of scale, or *network optimization*. There is no need to maintain duplicate networks: one for cell-to-MSC connectivity and one for interconnection (PSTN trunks-to-MSC) connectivity. Both cell base station radio trunks and PSTN interconnection trunks can traverse the same fixed network.

16.3   Transmission Media

A number of transmission media are available for use by wireless carriers when implementing their fixed networks. The most common media used are copper cable, optical fiber, and microwave radio systems.

16.3.1   Copper and Fiber

In leasing facilities from a telephone company or CLEC, the two prevailing options are copper cable and optical fiber. These options may be limited because of the availability of facilities. In rural areas, it is unlikely that fiber would be available. In some rural areas, especially in the northwest United States, copper cable is not even available for leasing from local telcos!

Optical fiber transmission facilities are now ubiquitous in metropolitan areas of the United States. When fiber *is* available in metro areas, it is usually in the form of a SONET ring, which provides for network redundancy. In this age of increasing competition, some new telephone companies (CLECs) are even willing to build fiber rings on a contract basis in exchange for a committed amount of traffic (over the rings) over a certain number of years. For example, one cellular carrier operating in the Raleigh-Durham area contracted with Time-Warner Cable to build a fiber ring for its fixed network. The main disadvantage to leasing transmission facilities from local and competitive telephone companies is that a wireless carrier is completely dependent on the telephone company in terms of network *reliability.* However, this disadvantage can be downplayed if the telco facilities are redundant (e.g., SONET rings).

16.4   Network Operation Centers

Like any long distance or local telephone company, all wireless carriers maintain some type of network operations center (NOC). NOCs are usually manned on a 24 × 7 basis (24 hours a day, 7 days a week, 365 days per year). Modern NOCs contain a large room with the status of various systems funneled onto a large, overhead projector screen. NOCs also have the Weather Channel on at all times to monitor weather across the country, as severe weather could have a major impact on system integrity.

In the wireless world, all MSCs are connected back to the NOC for purposes of monitoring the technical health of a carrier's system nationwide. Technically, this monitoring operates in a tiered manner, going all the way down to the base station level. All cells are connected to an MSC, and all MSCs are connected to the NOC.

A wireless company, like other telephone companies, can have one or more NOCs, depending on its size. If there is more than one NOC, they will be regionalized. They'll be geographically distant, but placed symmetrically so they can support entire regions of the United States. In cases where a carrier has multiple NOCs, the NOCs themselves will also be connected together as well to act as backups for each other.

The NOC also performs all of the core network management functions of any standard wide-area data network: fault management (system

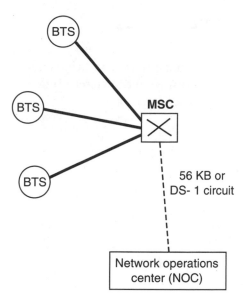

Figure 16-4
Network operations centers.

MSC

BTS

BTS

BTS

56 KB or
DS- 1 circuit

Network operations
center (NOC)

alarms), configuration management, security management, accounting management (traffic data), and performance management.

Many NOCs also house large data centers where the billing system for the carrier resides. The centers use completely redundant information systems (databases) known as *clusters,* where a mirrored copy of all billing data within the system is copied at given intervals from one system to another. In some cases, the data mirroring is done in real time. See Figure 16-4 for an illustration of the network topology that feeds into a network operations center.

16.5   Test Questions

True or False?

1. _____ A stand-alone link that runs off the main backbone of a daisy-chain network is known as a *spur.*

2. _____ The cellular fixed network cannot be used to transport both cellular radio (trunk group) traffic as well as interconnection (trunk group) traffic.

Multiple Choice

3. What are the three atmospheric conditions that have the most control over the microwave radio system index of refractivity?
   (a) The nightly weather forecast
   (b) Temperature
   (c) Sunlight
   (d) Humidity
   (e) Wind speed
   (f) Barometric pressure
   (g) a, b, and d only
   (h) b, d, and f only
   (i) d, e, and f only

4. The process of transporting cellular system traffic across the fixed network back to the MSC for switching is known as:
   (a) Demultiplexing
   (b) Downlinking
   (c) Cell relay
   (d) Backhauling
   (e) None of the above

CHAPTER 17

Microwave Radio Systems

17.1 Overview

Modern digital microwave radio systems provide a feasible technical solution for telecommunications transmission links at distances up to 80 km (approximately 48 mi) point to point. Planning and developing a microwave system in a wireless carrier environment is a dynamic and continuous process.

A typical microwave radio node ("terminal") consists of three main components: an indoor mounted baseband shelf, an indoor or outdoor mounted RF transceiver, and a parabolic antenna. Each node transmits and receives information to and from the opposite node simultaneously, providing full duplex operation.

Many wireless carriers choose not to deploy extensive microwave radio systems throughout their network. However, one cellular carrier, predominantly operating in rural America, has implemented one of the largest private microwave networks in the country for its fixed network. The main advantage to deploying a large, *private* fixed network (using microwave radio) is that a wireless carrier has ultimate *control* over that network in terms of the reliability of the system and the nature of the hardware components that are purchased and deployed.

The objective for any microwave system is to provide the best distortion-free and interference-free service. Overall, reliability of a microwave system depends on equipment failure rates, power failure rates, and propagation performance of any individual path.

Key: A (point-to-point) microwave radio system is sometimes referred to as a "hop" or a "shot."

17.2 Microwave System Development and Design

17.2.1 Network Documentation

At any given time, a design engineer should be able to view the transmission network's evolution. Therefore, a preliminary, documented network design is required in the form of a large diagram, to establish all the nodes in a network which require transmission links between them.

It can become the main reference document for network planning and implementation. Standard symbols should be adopted and agreed upon internally and should be strictly adhered to. This symbol structure should permit illustration of the different types of elements in the network, along with the varying capacities utilized on the different links. Once a network diagram is established, it can help to evaluate consequences of future network growth, and forecasts for the future can be superimposed accordingly onto the network map.

Although many modern digital microwave products are designed to be modular, providing "minimum disruption" upgrade routes, for a number of logistical reasons it is beneficial to have allowance for capacity growth inherent in the network from the outset. Future growth in the number of transmission links can have significant consequences in terms of site selection, and the network diagram can be used as a vehicle to highlight these areas. A new diagram should be produced every 3 to 6 months, as it is important to look at the consequences of growth both in terms of capacity and number of microwave links.

17.2.2   Network Design

The network design should be drawn to plan and illustrate the network topology to be adopted. Typically, there are two types of microwave network topologies in use, namely, star networks and ring networks. Such topologies will contain one or more hub sites at strategic locations that serve spurs or chains of subordinate sites from the centralized hub. In many cases today, the star and ring topologies will be combined, to form a hybrid star-ring topology. The hub sites in these networks should be limited to serving a maximum of six or seven cell sites to maintain good network reliability. The completed star-ring network design is accomplished by implementing transmission *loops*, or rings, between the hub sites in the network. It requires one extra transmission link from *each hub site* to two other hub sites. The advantage is that the rings can be used to provide *path diversity* and integrity to the network, removing the need for duplication of single links. See Figure 17-1 for a depiction of a hybrid star-ring network.

Key: Ring structures can be successfully achieved only if the necessary routing and grooming intelligence exists at all appropriate points in the network. Further, the capacity of each link in any one ring *has to be* sufficient to support all sites in the loop.

Figure 17-1
Transmission loops in
a microwave network
infrastructure provide
diverse routing,
increasing transport
system reliability.

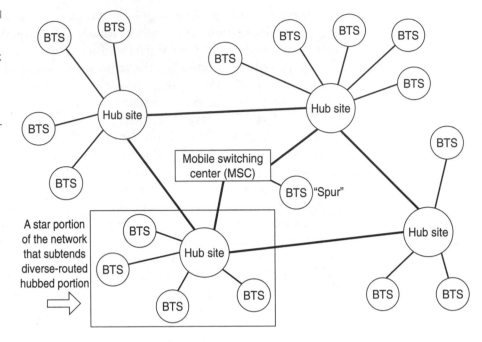

17.2.3 Site Selection

Once the prospective site is accepted by all functional groups (i.e., engineering, operations, real estate, regulatory), the final microwave design can be completed. This will entail a final path study, urban/rural area considerations, frequency selection, meeting with the regulators, a review of available frequency bands, frequency approval, and finally decisions involving weather and frequency band versus path distance considerations.

Minimizing the number of required sites in the network will bring logistical benefits and control real estate investments or site leasing costs. Therefore, it is always critical when selecting sites that no specific network element is considered in isolation. There are a number of specific items to bear in mind when designing microwave radio fixed links:

■ Good microwave sites, particularly in relation to hub sites, will be relatively high geographic points to give maximum line-of-sight availability. If at all possible, the wireless equipment and the microwave fixed link outdoor equipment should share any required towers or poles. Likewise, indoor microwave equipment can be housed in the same equipment shelters as the other radio equipment (i.e., cellular, PCS) and should be planned accordingly.

■ Required loading needs to be calculated if new tower installations are proposed, and the loading calculations must take antennas, wind, and ice loading into account. If a new microwave system is being added to an existing tower, calculations are required to ensure the incremental loading can be handled within the specifications of the existing tower.

■ If a site is in a rural or remote area, the service access should always be considered, particularly in times of inclement weather such as heavy snowfall.

■ Attention should be given to future growth requirements in all areas, especially if the site is likely to develop into a future hub. It is a good practice to inform landowners of any potential future growth to prevent problems at a later date.

17.2.4 Path Studies

A microwave path is the geographic span comprising the two ends of a link, the A end and the Z end. A final path study must include propagation analysis and take into account reflection surfaces such as lakes, rivers, drainage fields, sandy areas, marshland, and large flat roofs. All these things could significantly affect how RF in a microwave path will propagate from one end of the link to the other end.

If the microwave path is in an urban area and the corresponding site can be seen with binoculars or a telescope, this is generally sufficient for line-of-sight confirmation.

In rural areas the microwave paths are usually longer (greater than 9 mi) and use tower structures instead of buildings. It is difficult to verify the line of sight (LOS) if the tower is not yet built. In these instances the microwave path must be plotted on a map, and an analysis performed (using software tools) to determine antenna heights, taking into consideration clearance and reflection criteria.

17.2.4.1 Line of Sight It is fundamental to the correct performance of any microwave radio link that line of sight is available—that there is a clear transmission path between the two nodes of a link. This means that there can be no natural or man-made obstructions in the proposed path between the two ends of a microwave system. There are two ways of establishing line of sight: by creating a path profile, or by surveying the actual path physically.

A path profile is established from topographical maps which, by reference to the contours of the map, can be translated into an elevation profile of the land between the two sites in the path. Earth curvature can be added, as can known obstacles. The Fresnel zone calculation (see Section 17.2.5, "Fresnel Zones") can then be applied and an indication of any clearance problems can become known. (There are various software tools available to assist this exercise if required.)

A path survey can be undertaken by visiting sites and observing that the path is clear of obstruction. It is important to make note of potential future interruptions to the path such as tree or foliage growth, future building plans, nearby airports and subsequent fight-path traffic, and any other transient traffic considerations.

When transmitting from one end of a microwave path to the other, the electromagnetic signal disperses as it moves away from the source, and therefore the LOS clearance must take this dispersion into account. Particular attention should be paid to objects near the direct signal path, to ensure that the required signal levels reach the receiving antennas. This is referred to as the Fresnel zone clearance.

It is a matter of a particular wireless carrier's engineering practice which of the routes to establishing line of sight is utilized, and it will be dependent upon factors such as link length, site locations, availability of topographical information, and availability of tools. It is not uncommon to use both techniques—path profile and path survey—for certain links.

17.2.5 Fresnel Zones

Fresnel (pronounced frah-nell) zones are described as the route that reflected, or *indirect*, microwave radio energy takes to get to the distant-end microwave radio receiver. All the electromagnetic energy of a microwave radio path does not traverse a *direct* path between the transmitter and receiver. The Fresnel zone represents an ellipsoid under the direct radio beam. At the distant-end receiver, this *indirect* energy (the reflected signal path) either adds to or detracts from the energy of the direct radio beam between the two microwave antennas. From a graphic perspective, Fresnel zones represent a U-shaped line placed underneath the *direct* line of radio energy between the transmitter and receiver.

There are *even*-numbered Fresnel zones and *odd*-numbered Fresnel zones. They exist in layers underneath the direct signal. These different Fresnel zones are determined by the respective degree of phase reversal of the indirect radio signal that occurs along the route between the transmitter and the receiver.

Odd-numbered Fresnel zones incur a half-wavelength phase reversal (180°) between the transmitter and the receiver, but the indirect radio energy arrives at the receiver *in phase* with the direct radio signal. Therefore, odd Fresnel zones add to, or complement, the total, composite radio signal at the receiver because they arrive in phase with the direct signal. All odd Fresnel zones (i.e., first Fresnel, third Fresnel, etc.) are half-wavelength multiples of the direct radio beam. For example, first Fresnel is a half-wavelength phase reversal of the direct beam, and third Fresnel is a $1\frac{1}{2}$-wavelength phase reversal of the direct radio beam.

Even-numbered Fresnel zones also incur a half-wavelength phase reversal (180°) between the transmitter and the receiver, but the indirect radio energy arrives at the receiver *out of phase* with the direct radio signal. Therefore, even Fresnel zones lessen, or detract from, the total, composite radio signal because they arrive out of phase with the direct signal. All *even* Fresnel zones (i.e., second Fresnel, etc.) are full-wavelength multiples of the direct radio beam. For example, second Fresnel would be a one full wavelength phase reversal of the direct radio beam. Fourth Fresnel would be a two full wavelength phase reversal of the direct radio beam, and so on. See Figure 17-2.

Key: The goal in designing a microwave system is to ensure that no more than first Fresnel is obtained between the direct radio beam and the terrain in order to avoid unwanted signal reflections.

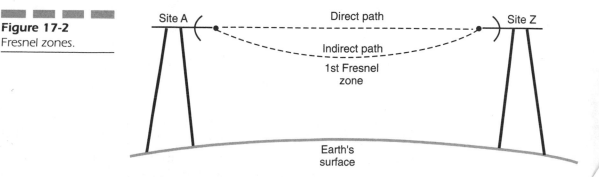

Figure 17-2
Fresnel zones.

17.2.6 Frequency Management

There are many factors that determine which frequency band will be used in a microwave system. Early microwave links were implemented using lower frequencies, such as 2 GHz and below. Frequencies were easily obtained and equipment was readily available. But these bands are now congested. Typical microwave frequency bands now in use in wireless networks are 8, 10, 13, 15, 18, 23, and 38 GHz, though some of these may not even be available in some countries. Wireless operators are usually assigned three or four frequency bands for the design of the fixed network, the most common being 8, 15, 18, and 23 GHz. It should be noted that in the United States, the 8-GHz band is today primarily used in private networks.

Weather is an important consideration when designing microwave systems, since it can affect their efficiency.

Key: Raindrops attenuate higher frequencies, and this must be taken into account when working with frequencies of 10 GHz and higher in microwave systems.

Along with rain outage, higher-frequency bands are attenuated more by the atmosphere (higher free-space loss) than lower-frequency bands. Optimally, frequency bands should be matched to a given microwave path as follows:

- Higher frequencies should be assigned to shorter paths.
- Lower frequencies should be assigned to longer paths. While the disadvantage of using higher-frequency bands on longer paths is rain outage, the drawback of using lower frequencies on shorter paths is frequency congestion. For example, if 15-GHz systems are used for path distances of 1 mi, this will use up the 15-GHz frequency band quickly, making it unavailable for future path distances of 4 to 9 mi. Practical design calls for using the 18- or 23-GHz band for 1-km paths and the 15-GHz band for longer paths. Table 17-1 shows the ideal path distance to frequency-band association in microwave networks.

Once the frequency band has been chosen, the proper frequency channel must be assigned to the microwave link. This should be selected so that it will not interfere with other operations systems (i.e., PCS base stations transmitting at 1.9 GHz in the United States).

TABLE 17-1	Path Distance	Ideal Frequency Band
Microwave Frequency Band and Path Use Recommendations	Less than 4 mi	18 or 23 GHz
	4 to 12 mi	13 or 15 GHz
	Greater than 12 mi	2, 8, or 10 GHz

With the preliminary network design in place, a clear picture is available of the different path lengths and capacities of links required. It is then necessary to determine if this is achievable within the terms of local regulations governing frequency availability and management. As stated previously, the propagation characteristics of electromagnetic waves dictate that the higher the frequency, the greater the free-space loss, or attenuation, due to the atmosphere. This means that frequency reuse distances are shorter: the distance between links operating on the same frequency can be shorter *without* fear of interference between these links! As a result, using lower-frequency bands for longer paths, and higher-frequency bands for shorter paths, as shown in Table 17-1, can make more efficient use of the frequency spectrum. The majority of national frequency management administrations will also have some form of link-length policy in adherence with this philosophy.

17.2.6.1 Microwave Frequency Bands   There are six frequency bands where private microwave operators can apply for FCC licenses to implement microwave systems. These frequencies are in the "common carrier" bands: L, S, X, Ku, K, and Kn bands. (See Figure 17-3.)

There are also small-capacity, unlicensed microwave systems that do not require licensing by the FCC. They most often use the 800-MHz frequency range, and they have a 1 to 4 DS1 capacity.

17.2.7   Diversity and Protection Systems

Microwave systems are available in nonprotected and protected configurations. Several protection schemes are available, including monitored hot standby (MHSB), frequency diversity, and space diversity.

From an equipment perspective, a protected terminal provides full duplication of all active elements, for example, both the RF transceiver and the baseband components. This is an example of MHSB.

Figure 17-3
Microwave frequency
bands.

FREQUENCY BANDS

HF	3-30 MHZ
VHF	30-300 MHZ
UHF	300-1000 MHZ
L-BAND	1.0 - 2.0 GHZ
S-BAND	2.0 - 4.0 GHZ
C-BAND	4.0 - 8.0 GHZ
X-Band	8.0 - 12.0 GHZ
K_U BAND	12.0 - 18.0 GHZ
K-BAND	18.0 - 27.0 GHZ
K_A BAND	27.0 - 40.0 GHZ
MILLIMETER	40 - 300 GHZ

Both space diversity and frequency diversity provide protection against path fading due to multipath propagation, in addition to providing protection against equipment failure. These techniques are typically only required in frequency bands below 10 GHz, specifically for long paths over flat terrain or over areas subject to atmospheric inversion layers (i.e., bodies of water or high-humidity areas).

Space diversity requires use of additional antennas, which must be separated *vertically* in line with engineering calculations. Frequency diversity can be achieved with one antenna per terminal, configured with a dual-pole feed. However, it should be noted that this will complicate frequency management in the long run.

If any microwave link carries traffic from more than one site, it needs some sort of protection mechanism. As the network's number of cell base stations increases, a system of transmission loops should be established between major hub sites and the switch (or switches) to increase survivability and reliability in the network. Transmission loops in a network infrastructure provide *diverse routing*, thereby increasing the transport system's reliability. One option is to keep a redundant route available with the capacity to carry *all* the traffic (of the loop) in case the main route fails. Another option is to use the diverse route at all times so that traffic flow is split 50-50 between the two routes. This design reflects what is known as *load balancing.* If unprotected radios must be used, they should be deployed only to serve single end sites, known as "spurs." See Figure 17-1 for a depiction of transmission loops in a wireless carrier environment.

17.2.8 Microwave System Capacity

Most microwave radio systems in use today employ digital carrier systems using time-division mutiplexing (TDM). There are very few (if any) analog microwave systems existing in the field today. The older analog systems used frequency-division multiplexing (FDM). Some utilities may still be using the older analog microwave systems.

Capacity of a microwave system is an important consideration. Microwave radios can be configured to carry a certain amount of traffic in a specific frequency range. Capacities range from DS1 all the way up to OC3 (three DS3s).

Example: A carrier could select a 16-DS1 capacity radio operating at 23 GHz to carry a significant amount of traffic over a path distance of 4 mi. A carrier could also select a 12-DS1 capacity radio operating at 2 GHz to carry traffic over a path distance of about 23 mi.

17.2.9 Microwave System Reliability: Index of Refractivity

The reliability of the propagation of a microwave radio path is impacted by what is known as the *index of refractivity* of the microwave radio beam. Three atmospheric conditions have the most effect on the index of refractivity: barometric pressure, humidity, and temperature.

Temperature inversion is a condition that has an adverse impact on microwave radio systems, and may cause the systems to become inoperative at times. About 99% of the time, the air closer to the ground (and closer to the microwave radio system) is warmer than the air in the lower atmosphere. When the opposite occurs and the air closer to the ground is cooler than the air high above—usually because of weather fronts—temperature inversion occurs. Temperature inversion distorts the microwave radio beam. As a rule, the microwave beam normally bends *toward* the earth. When temperature inversion occurs, the beam bends *away* from the earth, causing the microwave system to become inoperative. Temperature inversion occurs more frequently at higher radio frequencies, especially in the 11/18-GHz frequency range.

17.3   Coax and Waveguide

There are two choices of cable available for use with microwave antennas: either coaxial cable or waveguide. Coax cable is usually used for microwave radios that employ frequencies of 3 GHz or less. Waveguide is a hollow, elliptical metal cable that connects the RF equipment to the microwave antenna. Waveguide cable can be much heavier than coax cable, and its weight must be taken into account carefully for tower loading purposes.

17.4   Microwave Radio Antennas

Today's microwave antennas come in two predominant forms. The most common form is a type that appears as a concave dish on communication towers: the parabolic reflector.

17.4.1   Parabolic Reflectors

These are the most common type of microwave radio antennas in the common carrier band today. They appear as dishes on towers, and the dishes come in diameters of either 6, 8, or 10 ft. This type of antenna has a feed horn mounted inside of the dish. Both the transmit and receive signals are reflected off the dish into and out of the feed horn. About 90% of new microwave systems use this type of antenna. See Figure 17-4.

Radomes are dome-shaped or flat-shaped fiberglass covers for parabolic microwave antenna dishes. Their main purpose is to reduce wind resistance and thus tower loading as well. Radomes also protect the antenna itself from the elements.

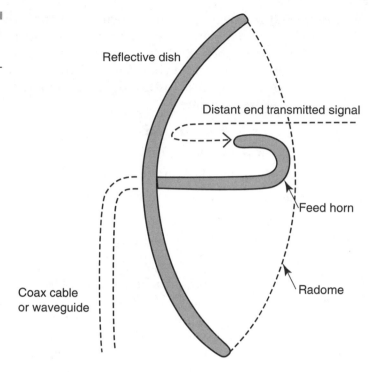

Figure 17-4

Parabolic reflector
microwave antenna.

Reflective dish

Distant end transmitted signal

Feed horn

Radome

Coax cable
or waveguide

17.4.2   Horn Reflectors

This type of antenna is similar to the parabolic reflector. Radio energy is beamed off a reflector inside the horn itself. In effect, the radio energy is funneled out of the horn. This antenna type has a direct radio feed by waveguide, as opposed to the parabolic reflector antenna, where the feed horn reflects the radio energy into and out of the system via the dish.

17.5   Microwave Radio System Software Modeling Tools

Microwave radio engineering can be as complicated and specialized as radio-frequency engineering for cell sites. As with RF engineering, there are special software programs that microwave engineers can use to determine the feasibility of installing microwave radio systems, and whether line of sight even exists in the first place between the intended transmitter and receiver locations. When using the software programs, a microwave radio engineer can change many of the parameters of a simulated microwave radio path in order to determine what factors, together, would optimize the radio shot. Some of the parameters (of simulated microwave radio systems) that can be changed to determine feasibility of a specific path are as follows:

- The microwave radio frequency
- The size of the antenna (dish)
- The height of the dish on the tower
- The diameter of the cable (if coax is being used)
- The transmitter output power
- The receiver threshold (the sensitivity of the receiver, or how well it accepts the transmitted signal)

Two of the more prevalent microwave radio software modeling programs are *Rocky Mountain* and *Pathloss.*

The reliability of a microwave radio system is a company-specific issue, and depends on the actual grade of service that the carrier wants to provide. However, most companies engineer microwave radio systems for a reliability factor of **99.99%**. This usually equates to only several minutes of outage *per year.*

17.6 FCC Form 494—Microwave Radio Systems

As stated in Chapter 11, "Cellular Regulatory Processes," the 494 form must be filed with the FCC to obtain permission to activate a microwave radio system.

17.7 Test Questions

True or False?

1. _____ There is more rapid fading of microwave signals at higher frequencies.

2. _____ The ultimate effect of Fresnel zones in microwave radio systems is that, at the receiving end, they either add to (complement) or detract from (cancel out) the direct signal between the A site and the Z site.

Short Answer

3. Name and discuss the key points of microwave path design.

Multiple Choice

4. Even-numbered Fresnel zones:
   (a) Will arrive at the distant-end receiver out of phase with the direct radio beam. Thus, their impact will detract from the total, composite signal arriving at the receiver.
   (b) Will arrive at the distant-end receiver in phase with the direct radio beam. Thus, their impact will be to supplement the total composite signal arriving at the receiver.
   (c) Will have no effect on the composite signal at the distant-end receiver.

5. The oval- or flat-shaped fiberglass covers used with parabolic reflector microwave antennas are known as:
   (a) Wind domes
   (b) Enhancers
   (c) Equalizers
   (d) Radomes

18

Wireless Interconnection to the Public Switched Telephone Network

18.1 Overview

Interconnection is defined as the point where any two (carriers') networks are linked together. It can involve a wireless carrier's connection to a local exchange carrier (LEC, a landline telco), or an Internet service provider's (ISP's) connection to an LEC, or an LEC's connection into an interexchange carrier (IXC).

Approximately 75% of all cellular telephone calls made today are initiated from mobile/portable phones, with the destination being a landline telephone. This number has been slowly but steadily changing to the point where more traffic is being terminated on wireless carriers' phones. These types of calls are known in the wireless industry as mobile-originated, or *mobile-to-land* (M-L) telephone calls. For example, a driver returning from work on the expressway may call home to check on the family. Because of this large volume of mobile-land wireless traffic, it is imperative that a wireless carrier develop and maintain a cost-effective series of interconnections to the public switched telephone network (PSTN). The public switched telephone network is that which we use every day to place our telephone calls, whether they are across town, cross-country, or across the world. In its most basic form, the PSTN is accessed every time a person picks up a telephone and hears a dial tone.

To develop the interconnections, a wireless carrier must first examine *where* to place interconnections and what *type* of interconnections should be ordered. This decision can be based on several factors.

First, a logical location for connection to the public telephone network is to place an interconnection, also known as a *point of presence* (POP), where the competition has an interconnection. That way a wireless carrier can stay abreast of the competition, in terms of providing "local" service and a local telephone number where the competition is doing the same. (The term POP actually refers to a point where any two telecommunication carriers link together, whether they are local telephone companies, long distance carriers, etc.)

Second, cost factors must be taken into account to justify the placement of an interconnection. Marketing projections should prove that the forecasted amount of sales (and in turn, call volume) will outweigh the installation and monthly charges that are inherent to interconnection. Also, in some cases special construction charges may apply to install interconnection circuits, especially in rural areas where a wireless carrier's cell site may be miles away from the local telephone company's nearest cable terminal or central office. These charges can range in price anywhere from $3000 to $50,000 or more!

Third, the population density of a given geographic region must be examined to determine whether a local interconnection will suffice, or whether a connection that offers low cost-per-minute charges over a wider area is more appropriate. These issues will be explained in more detail later.

There are many facets to interconnection: the elements which compose the interconnection itself, the different types of interconnection available, the cost structures, and legal issues relating to how interconnection agreements are formalized. These will all be explained in this chapter.

18.2   The Structure of the Public Switched Telephone Network

18.2.1   The Breakup of AT&T and Current Makeup of the PSTN

In 1983, Judge Harold Greene issued a consent decree from U.S. District court in Washington, D.C., that broke up the Bell telephone system, which was monopolized by AT&T up to that point in time. The breakup is also commonly known as the Modified Final Judgment (MFJ). AT&T was required to divest itself of ownership of the infrastructure and assets of the majority of the public switched telephone network. With the breakup of AT&T in 1983, seven regional Bell operating companies (RBOCs) were created. The RBOCs are also known as the "Baby Bells." All seven original companies covered contiguous geographic sections of the United States. Most of the seven RBOCs were named according to the geographic region of the United States where they offer service: Ameritech, Pacific Telesis, U.S. West, Nynex, BellSouth, Bell Atlantic, and Southwestern Bell. From 1995 to 1999, most of the Bell companies have merged. Bell Atlantic has merged with Nynex and GTE, and is now renamed "Verizon Communications." Southwestern Bell, now known as SBC Communications, merged with Pacific Bell in 1997. Then in 1999, SBC bought Ameritech, thereby reducing the total number of incumbent Bell companies to four: SBC, Verizon, BellSouth, and U.S. West. It should be noted that U.S. West and Global Crossing also merged in 1998.

The RBOCs are the *dominant* local telephone companies in their respective regions. The local telephone company, anywhere, is known as the local exchange carrier (LEC), or the "telco." The remainder of the local telephone company landscape, nationwide, is composed of independent telephone companies. Some of these independent companies are very large, such as the Sprint local telephone companies and the former GTE. These larger independent telcos have distributed footprints throughout the United States. Others are "mom and pop" telcos, usually named according to the geographic areas where they operate. Examples are Lufkin-Conroe Telephone Company of Lufkin, Texas, and Mud Lake Telephone Company in Idaho.

As the RBOCs are the dominant carriers in their states, they are more heavily regulated than the independent telephone companies. This is a result of a regulatory mandate of ensuring that there is a "level playing field" in terms of the competitive landscape between CLECs, the independent telcos, and the RBOCs.

When a wireless carrier obtains interconnection to the public network, it is obtained from either an RBOC, a CLEC, or an independent telephone company. This depends on which type of phone company has been granted the right to operate in a specific geographic area by the respective state Public Utility Commission.

18.2.2 Local Access and Transport Areas

When AT&T was divested in 1983, there was more to the breakup than dividing the country up into seven regional Bell companies. As part of the divestiture, the MFJ called for the separation of "exchange" and "interexchange" telecommunications functions. What this meant was that exchange services were provided by the regional Bell operating companies or independent telephone companies, and interexchange services were to be provided by interexchange carriers. The United States was thus broken up into 197 geographical areas known as local access and transport areas (LATAs). LATAs serve the following two basic purposes:

> They provide a method for delineating the area within which the LECs may offer services.

- They provided a basis for determining how the assets of the former Bell System were to be divided between the RBOCs and AT&T at divestiture (i.e., outside and inside plant).

As the name implies, the basic principle underlying the structure of LATAs is that *within* a LATA, only local telephone companies (LECs) could provide exchange services (carry local telephone traffic). LATA sizes vary and are usually based on population densities of a given area. For example, in the western United States, because of the sparse population of some states, an entire state may be one LATA. In other states such as Illinois, there are about a dozen or more LATAs. Each LATA has a 3-digit number assigned to it for identification purposes. For example, Chicago, Illinois, is in LATA number 358. In LATAs where an independent telco is the dominant local exchange carrier, the LATA designation number begins with the number 9. For instance, LATA number 949 is in North Carolina, and the dominant LEC there is the independent telco Sprint Mid-Atlantic Telephone Company. For the most part, LATAs do not encompass more than one state in the United States, although they can cross state boundaries in some cases.

LATAs were designed not only to delineate exchange areas, but also to keep intact geographic areas that were socially, economically, and culturally tied together. This is why Chicagoland and Northwest Indiana—in two different states—remained one LATA (#358). These two areas are tied very closely together in the aforementioned ways.

18.2.3   The Telecom Act of 1996

From 1983 to 1996, any traffic that crossed LATA boundaries had to be carried by an interexchange carrier (IXC). An interexchange carrier, also known as a long distance carrier, is a telephone company that carries long distance traffic from one LATA to another LATA (inter-LATA), from state to state (interstate), and internationally. However, from 1983 to 1996, IECs were *forbidden* to carry traffic anywhere *within* LATAs (intra-LATA). But in some cases IXCs received approval to do so on a state-by-state basis. Simultaneously, the local telephone companies were *forbidden* to carry traffic from one LATA to another (inter-LATA).

In the Spring of 1996, President Bill Clinton signed the Telecommunications Act of 1996 (TA '96) into law. The ultimate intent of TA '96 is to open all telecommunications markets to competition. This includes local exchange service, intra-LATA toll service, inter-LATA/interstate long distance service, and cable television service. Landline carriers of all types will eventually be allowed into each other's markets. Just as the *wireless* world will have intense competition over time due to the entrance of PCS companies into the market, the same onset of competition will eventually apply to the landline carriers. Because of this impending competition, service offerings by landline companies should become cheaper and more expansive. This will supposedly benefit wireless carriers in the form of offering multiple, less-expensive options for interconnection to the PSTN. The only caveat for wireless carriers when choosing among a number of companies offering local and long distance interconnection is to ensure that the companies offering interconnection services provide *reliable, quality service*. Because of this caveat, it will behoove wireless carriers to ensure they do *not* sign any long-term contracts for interconnection services for several reasons:

> If it turns out that a certain carrier is not offering quality service, a wireless carrier would want to ensure that they are not tied down to a long-term contract, so they would not have to pay a termination liability to the landline carrier for terminating interconnection contracts. The wireless carriers should ensure that there is a reasonable escape clause in their contracts.

■ Because there will likely be multiple carriers offering interconnection, a wireless carrier may want to retain the option to keep trying to interconnect with *various* PSTN companies until they find a company that offers quality, low-cost interconnection services.

18.3   Telcordia Reference for Interconnection

The technical reference for wireless interconnection to the PSTN is found in Telcordia General Reference (GR) 000-145. This document provides the technical definitions associated with any type of interconnection between two service carriers.

18.4   Elements of Interconnection

There are three basic elements to wireless interconnection to the PSTN: transport via transmission systems (DS1, DS3, OC-*n*), trunks, and telephone numbers. Each DS0 acts as a trunk in this context. In the majority of cases today, telephone numbers are assigned to wireless carriers (through the telephone companies) free of charge, through a company called NeuStar, the North American Numbering Plan Administrator (NANPA). Some independent telephone companies still charge for the numbers, but this is becoming more and more rare. Just several years ago, most telcos charged wireless carriers when they assigned them ranges of telephone numbers. The highest charges—in thousands of dollars—applied when wireless carriers obtained entire NXX codes.

18.4.1   The DS1 Circuit

The DS1 circuit is a 4-wire digital circuit composed of 24 channels, or "time slots." The DS1 also goes by many other names, such as T1, T-span, high-cap, and pipe, to name a few. The overall speed of a DS1 circuit is 1.544 megabits per second (Mbps). This speed equates to 64 kilobits per second (kbps) per channel plus 8 kbps for timing and framing for the entire DS1 circuit. Two wires are used to transmit, and two wires are used to receive.

Each channel (DS0) is capable of handling one telephone conversation at a time. The modulation technique which is employed with a DS1 circuit is time-division multiplexing (TDM). With time-division multiplexing, each channel is assigned a particular moment in time with which to transmit information. Since there are 24 channels on every DS1 circuit, the maximum capacity of a DS1 is 24 simultaneous conversations.

18.4.2   The Trunk

It is important to understand the difference between a trunk and a line. A line is a circuit which connects a piece of equipment to a switching system or a computer system. For example, a person's home telephone is connected to the telephone company central office by a line. A trunk connects two switching systems together, or a switching system to a computer system. For example, in the case of wireless interconnection, DS0s running over a DS1 circuit function as trunks that connect the MSC with the switching system that resides at a telephone company central office. Each DS1 channel functions as a separate trunk.

18.5   Interconnection Operations

Interconnection is still a large expense for most wireless carriers, even though the cost-per-minute rates charged by the landline telcos have trended downward since 1996. It's very important for wireless carriers to ensure that they're always implementing the best type of interconnection throughout their markets. Additionally, it is wise to employ a small staff of billing analysts who can decipher the interconnection bills to determine if incorrect rates are being charged to the wireless carrier. In most instances, the staff of billing analysts can pay for their own salaries many times over by catching costly errors on interconnection invoices.

Calls originating from a landline telephone to a wireless subscriber (land-to-mobile calls) are routed through the PSTN to the telco central office where the wireless carrier's interconnection was originally obtained. Land-to-mobile calls are then routed from the telco central office to the wireless carrier's MSC over the actual interconnection links. The actual *switching* of wireless calls takes place at the MSC.

Even though the wireless carrier owns the rights to the numbers it obtains through interconnection to the PSTN, and switches them from the wireless switch, the telco switches know how to route incoming calls dialed from PSTN customers. This is because the issuing authority for telephone numbers, NeuStar, will send out a notification to *all* telephone companies in the United States (including CLECs and independent telcos) notifying them when a new line range or NXX code is activated and of the central office (code) where the line range is based. The telco stores the line ranges owned by the wireless carrier in the switch that connects the wireless carrier to the telco—the telco switch where the wireless carrier obtained interconnection. See Figure 18-1.

Wireless interconnections to the PSTN can occur from either a cell site location or an MSC location. If a desired interconnection can be accomplished from a cell site that is closer to a telco POP than the MSC, it makes more sense from a cost perspective to place the interconnection into the cell site. The interconnection traffic is then backhauled to the MSC for switching.

Figure 18-1
Elements of interconnection.

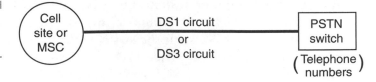

18.6   Types of Interconnection

There are multiple types of interconnection available to wireless carriers. Each type of interconnection has its own functional characteristics, and each type is also ordered for specific marketing and operational reasons. The underlying principle in ordering any type of interconnection used for carrying wireless traffic is to obtain the largest available geographic "footprint" at each point of interconnection for the lowest cost per minute across that footprint. The larger the footprint, the more cost-effective the interconnection. Each type of interconnection to the PSTN also has its own calling area. This will be explained in more detail below.

18.6.1   Type 2A Interconnection

A Type 2A interconnection consists of a trunk group between a wireless carrier's cell *or* MSC, to a telco access tandem, known in traditional Bell nomenclature as a Class 4 central office. Tandems serve as hubbing centers for two or more end offices in a LATA. An end office is also known as a central office (CO) or a Class 5 central office in traditional Bell nomenclature. An example of an end office is the type of central office that offers regular home or business dial tone.

Key: The end office is the lowest link in the chain of switching systems that comprise the PSTN.

Tandems exist in the PSTN hierarchy because it is not always practical or cost-effective for a local exchange carrier to link every end office in a given LATA to every other end office within that LATA. With (access) tandems in place, PSTN switching can become a hierarchical operation, which is more efficient than linking all end offices to each other. This also means that tandems will have high-capacity trunks connecting them all together, especially for purposes of linking geographically distant end offices. In some instances, the "local calling area" for Type 2A interconnections is the entire LATA. It is actually the objective of wireless carrier network engineers to obtain the largest possible LATA-wide footprint when obtaining Type 2A interconnections. Wireless carriers obtain an entire exchange code when procuring a Type 2A interconnection to the PSTN (i.e., 607-689-0000-9999) through the NANPA, NeuStar.

Mobile-terminated calls are routed to the tandem where interconnection to the wireless carrier takes place. Again, the NXX code resides in the MSC, not in the telco switch. Routing of calls through the PSTN *to* the wireless network is based on the mobile's NPA/NXX code, and the actual switching of wireless calls takes place at the MSC. (NPA stands for *numbering plan area,* or area code.)

Key: Type 2A interconnection represents the procurement of interconnection from public, landline telephone companies at a *wholesale* rate. The MSC is treated as another telco central office switch in terms of call handling and tariffs.

See Figure 18-2 for a graphical view of a Type 2A interconnection.

18.6.2   Type 2T Interconnection

A Type 2T interconnection is a trunk group to a telco access tandem that is used to route *equal access* traffic from a wireless carrier's network to an interexchange carrier's (IXC's) POP. Equal access provides wireless customers the opportunity to presubscribe to a given interexchange carrier. Type 2T interconnection was required of RBOC wireless carriers (i.e., Bell Atlantic Mobile Systems, Ameritech Wireless) until recent FCC rulings exempted them of this obligation. This requirement originally stemmed from the Modified Final Judgment issued in 1983.

This type of interconnection provides wireless customers access to all long distance carriers via a Feature Group D (FGD) trunk group, which carries the automatic number identification (ANI) of the calling party (wireless subscriber); the carrier identification code (CIC) of the appropriate IXC (10-XXX), and the dialed number (information digits). See Figure 18-3 for a graphical depiction of a Type 2T interconnection.

18.6.3   Type 1 Interconnection

A Type 1 interconnection consists of a trunk group between a wireless carrier's cell *or* MSC to a local exchange carrier (LEC) end office. This interconnection handles more functions than any other type of interconnection.

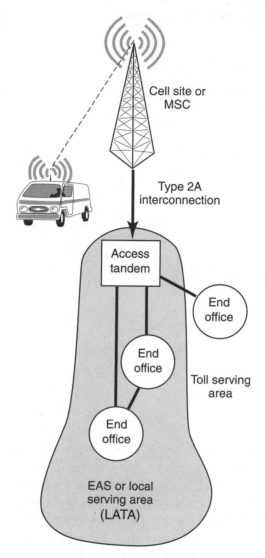

Figure 18-2
Type 2A interconnec-
tion.

Cell site or
MSC

Type 2A
interconnection

Access
tandem

End
office

End
office

End
office

Toll serving
area

EAS or local
serving area
(LATA)

This interconnection could be ordered for marketing purposes, simply to obtain local access telephone numbers, or a "local presence" in a
specific geographic area. This strategy applies especially in rural areas.
With Type 1 interconnections, the line ranges utilized by the wireless
carrier to serve its subscribers are actually "owned" by the telco and
assigned to the wireless carrier. Like other types of interconnection, the
actual *switching* of wireless calls takes place at the MSC.

Figure 18-3
Type 2T intercon-
nection.

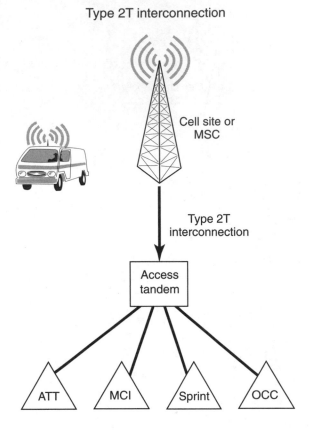

Type 2T interconnection

Cell site or
MSC

Type 2T
interconnection

Access
tandem

ATT   MCI   Sprint   OCC

In some rural areas, it might be necessary to order a Type 1 because it is not cost-effective to order other types of interconnection, such as Type 2A. For example, in a very remote but relatively large community, a wireless carrier would definitely want to order local telephone numbers. That way, calls from mobile phones to landline telephones (and vice versa) would be carried and rated at the cheapest possible rate, while simultaneously appearing as "local" numbers to the local population.

As a rule, calls that terminate outside the local calling area of Type 1 interconnections are rated as "toll" traffic by the local telephone company, and might therefore be rated as toll traffic by the wireless carrier as well. Sometimes wireless carriers will assume payment for these toll charges themselves, and not pass the charges on to their customers. This practice is known as "reverse billing" and is done for marketing and competitive purposes.

Along with local traffic, Type 1 interconnections also carry 911 traffic, 411 traffic, 611 traffic to wireless customer service centers, 0+

traffic (credit card calls), 0– traffic (calls to the operator), and 800 number traffic. Type 1 interconnections are also capable of carrying traffic to any destination LATA-wide, but Type 2A interconnections are the preferred method for transporting these calls because the rates are usually cheaper to terminate a call to a landline telephone via a Type 2A interconnection. 911 calls from wireless customers will be routed to a public service access point (PSAP) for processing.

Type 1 interconnections can also be used as a pipeline for traffic that needs to be routed to an interexchange carrier when the destination of the call is inter-LATA or interstate. When the amount of traffic in a given wireless market is not great enough to justify a *dedicated* DS1 circuit that ties the wireless carrier directly to an interexchange carrier's POP, wireless traffic is routed to the IXC over a Type 1 interconnection in switched access (SWAC) form.

Directory assistance (DA) 411 traffic can be routed directly to a telephone company call center over a Type 1 interconnection. Today, many wireless carriers are routing 411 traffic to independent call center companies over dedicated trunks from the MSC directly to the independent call center (ICC). The appeal of this service architecture is that the ICC will charge the wireless carrier a "wholesale" rate per call that is cheaper than what the local telco will charge. The wireless carrier will then charge their subscribers at a "retail" rate. This scenario results in a higher profit margin—per call—for the wireless carrier, and represents a viable revenue stream as well.

In the case of Type 1 interconnections, the NXX codes served out of the end office where the Type 1 interconnection is obtained are a reflection of what is known as the "local calling area," also known as the "extended area of service" (EAS). The local calling area itself reflects a concept known as the "community of interest," In most instances, the community of interest is an area made up of towns that are socially, economically, and culturally tied together. The community of interest was also a key factor that determined how LATAs were delineated when AT&T was broken up in 1983. An example of a community of interest is Chicagoland and Northwest Indiana. Another example would be two or three rural communities that have strong ties and are all served out of the same end office. The bottom line is that calls that are originated and terminated within the same EAS (local calling area) are always going to be rated cheaper than other calls. This is because these calls incur no transport; they are "intraswitch" calls.

Telephone numbers are obtained with Type 1 interconnections in blocks of 100 or 1000, depending on the telco.

Figure 18-4
Type 1 interconnec-
tion to the PSTN.

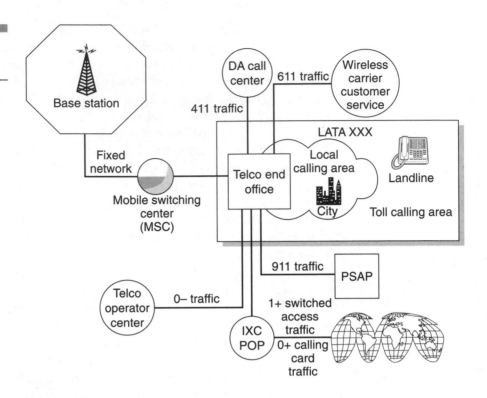

Key: Type 1 interconnection represents the procurement of interconnection from public, landline telephone companies at a *retail* rate. The
MSC is treated as a PBX in terms of call handling and tariffs.

See Figure 18-4 for a graphical view of a Type 1 architecture.

18.6.4 Type 2B Interconnection

A Type 2B interconnection consists of a trunk group between a wireless carrier's cell *or* MSC to a specific local exchange carrier end office.
These interconnections are solely used for local, cost-effective connection to the PSTN to terminate very high volumes of traffic to specific
exchange codes. This type of interconnection is justified based on wireless carrier traffic studies that show very high volumes of mobile-originated traffic terminating to several exchange codes that are all served
out of one end office. Although the Type 2B interconnection looks very
similar to a Type 1 interconnection, it is used only to terminate local

traffic to a specific CO. Besides the fact that no other types of traffic terminate over a Type 2B, such as 911 or 800 number traffic, the key differentiator with Type 2B interconnections is the fact that the cost per minute charged to the wireless carrier is less than it is with a Type 1 interconnection. Put simply, Type 2B interconnections are used for least-cost routing purposes. This type of interconnection is becoming more and more prevalent because there is no tandem-switching element involved in the rates charged to wireless carriers.

18.6.5 Dedicated Interconnection to an Interexchange Carrier (IXC)

If the volume of mobile-originated long distance traffic in a particular market is low, a wireless carrier would send all *inter-LATA* and *interstate* long distance traffic out a Type 1 interconnection as SWAC long distance traffic. In this scenario, the local exchange carrier routes long distance calls to the wireless carrier's preferred IXC on a *call-by-call basis*—through the LEC end office to the IXC's POP.

If the amount of long distance traffic in a wireless market becomes very large over a period of time, the wireless carrier can justify the installation of a dedicated interconnection to its long distance carrier. (See Figure 18-5.) This type of interconnection is justified by obtaining a quote for the monthly cost of a dedicated circuit from the wireless carrier's network, to the IXC's nearest POP. If the quote for its link is equal to or less than the monthly cost of sending switched access traffic to the IXC over a Type 1 interconnection (number of switched access minutes of use × cost per minute), the interconnection is justified. The rationale

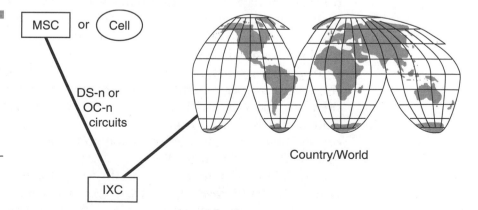

Figure 18-5
Dedicated connection to long distance carrier. Installation of these circuits is justified on the basis of volume of long distance traffic out of a particular MSC.

supporting installation of this type of interconnection is that it offers wireless carriers lower cost-per-minute rates versus sending traffic to the IXC through the LEC, via switched access. This is because access charges—in terms of higher cost per minute of traffic—are assessed to the IXCs by the LECs when long distance traffic is sent to the IXC through LEC central offices (over Type 1 interconnections). These additional costs—which may be up to 3 cents per minute—are then passed on to the wireless carriers. In turn, the wireless carriers will pass these costs on to their customers. In summary, this type of interconnection is a win-win situation for both carriers and their subscribers.

18.6.6   Intramarket Mobile-to-Mobile Interconnection

In some cases, it makes sense to do a direct, wireless carrier-to-wireless carrier interconnection *within* a wireless market.

Wireless carriers can obtain records from their switch (MSC) that will tell them how much customer-originated traffic in a given market terminates to given NXX codes in a market. A database provided by Telcordia known as the *Local Exchange Routing Guide* (LERG) lists all exchange codes within each NPA (area code). In the LERG, there is a column next to each exchange code which denotes the carrier who owns that exchange code. So through a combination of the switch records and the LERG, a wireless carrier can determine how much of its mobile-originated traffic terminates to its in-market competitor's telephone numbers.

Through nondisclosure and confidentiality agreements, two wireless carriers can determine how much of their respective customer-originated traffic terminates to the other wireless carrier's NXX codes (mobile-to-mobile calls) in that market. These carriers can then get a quote for the cost of a point-to-point circuit that would link their two networks (MSC to MSC). If the total cost to send the traffic to each other's NXX codes through interconnections to the PSTN (LEC) (X number of minutes times the cost per minute) exceeds the cost of the point-to-point DS1, then it would behoove these two carriers to implement a direct connection between their two networks to transport traffic between each other's subscribers. The carriers would split the monthly cost of the circuit evenly between themselves.

The intercarrier link is essentially used to support intermachine trunks (IMTs) between the two carriers' networks. These trunks then support customer-originated mobile-to-mobile traffic between the two carriers. The capacity of the facility connecting the two carriers' MSCs would be traffic-engineered based on the volume of calls between their subscribers. This is the key information that's obtained from the switch records that each carrier obtains, which spurred them to pursue this interconnection in the first place.

The key benefit of this type of interconnection is that the wireless carriers are essentially bypassing the LEC and saving substantial money on interconnection (to the PSTN) costs.

18.6.7   Point-to-Point Circuits

Point-to-point circuits are used to link two wireless carrier locations together. In the wireless world, this usually means linking two cell sites together: a cell site to the MSC, or business offices to MSCs. These types of circuits are installed when a wireless carrier does not have other means to link two locations together.

One example of this situation is where there is no line of sight for a microwave radio connection between two locations. This means that there is a natural or man-made obstruction between the two points which would block the signal of a microwave radio system. This could be either a mountain, mountainous terrain, or a building in a given area between the two locations which rises to a point where it blocks the radio signal. Another situation requiring a point-to-point circuit is when a cell site is so far away from another cell that a microwave radio link is not an option because the range would be too great for microwave radio.

Which type of landline carrier the wireless carrier approaches to provide a point-to-point circuit depends on whether the link between the two locations is intra-LATA or inter-LATA. If the two locations are intra-LATA, the wireless carrier approaches the local carrier. If the locations are inter-LATA, the wireless carrier approaches an interexchange carrier. The market for point-to-point circuits will ultimately be open due to passage of the 1996 Telecom Act. When a wireless carrier implements point-to-point circuits, the two ends of the circuit are known as the A and Z locations.

When the locations are in the same LATA (intra-LATA), the circuit passes through a local exchange carrier central office at both ends before arriving at its final destination. The part of the circuit between the wireless carrier's network (cell site or MSC) location and the local exchange carrier central office is known as the *local loop*. See Figure 18-6.

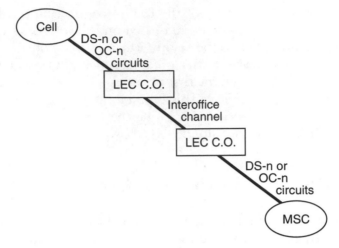

Figure 18-6
Intra-LATA point-to-point circuit structure.

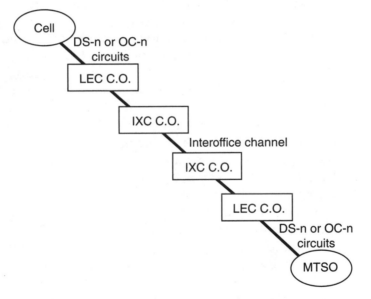

Figure 18-7
Inter-LATA point-to-point circuit structure.

When the locations are in two different LATAs, the connections *at each end* are made first to a local exchange carrier end office, and then to an interexchange carrier. The part of the circuit that runs between the interexchange carrier's central offices is known as the interoffice channel (IOC). When ordering inter-LATA point-to-point circuits through an interexchange carrier, there are times when the costs for the interoffice channel will be very high. This is because regardless of what the wireless carrier's A and Z locations are, the locations of some interexchange carrier central offices at each end could be extremely distant from the two locations the cellular carrier is trying to link together. If the circuit run is inter-LATA, the wireless carrier has no choice but to go through an interexchange carrier for service. The end result is that the total cost for the circuit may be very expensive, but if the wireless carrier needs the circuit, the cost may need to be overlooked. See Figure 18-7.

18.7   Cost Structures and Rate Elements

Most of the types of interconnection that have been reviewed in this chapter have some common cost structures and rate elements, as described below.

- Each interconnection is achieved by using a circuit to connect locations between two networks, the wireless carrier's network and the landline network, regardless of the carriers involved. For each circuit that is installed, there will be a nonrecurring charge (NRC) to install the circuit itself.

- For each circuit installed, there will be a monthly recurring charge (MRC) for the circuit itself.

- For certain types of interconnection, there may be a charge to activate elements of the circuit that act as trunks. This would be an NRC when it exists, based on the telephone company involved. An example of this activity would be the charge to activate DS0s for a DS1-based interconnection or the charge to activate DS1s for a DS3-based interconnection. For Type 2A interconnections, this charge no longer applies. It may apply for Type 1, Type 2B, or Type 2T interconnections.

- There may also be a recurring monthly charge for the trunks running over an interconnection circuit, based on the information above.

For point-to-point circuits, only two charges apply:

- An NRC for installation of the circuit
- An MRC for use of the circuit

Since point-to-point circuits are purchased as dedicated circuits, not circuits that have the telco switching traffic of any kind, no trunk charges apply for these types of circuits. Even if wireless traffic *does* get switched or routed over these circuits, when they're purchased by a wireless carrier, that carrier can use the traffic for whatever purpose it desires.

18.7.1 Cost-per-Minute Charges

Prior to the passage of the Telecom Act of 1996, rates for all types of interconnection ranged anywhere from 2 cents per minute (i.e., Type 2B interconnections) to more than 5 cents per minute (i.e., Type 2A interconnections).

Since passage of the Telecom Act of 1996, rates for all types of interconnection have dropped to a range of less than 1 cent per minute, to 3 cents per minute. Independent telephone companies usually charge higher rates per minute, because they have to compensate for the fact that they have much less volume to work with than, say, the RBOCs.

For dedicated wholesale interconnection to interexchange carriers, wireless carriers can obtain rates from around 3 to 6 cents per minute. The core issue, which drives the exact rate per minute charged to wireless carriers for this type of interconnection, is total billable minutes per year. At the same time, wireless carriers are usually required to sign long-term contracts of up to 3 years in order to obtain lower rates. Billing plans in today's retail wireless environment are often trumpeting the fact that customers will get free long distance. This reflects the fact that long distance service, in both wholesale and retail forms, is rapidly becoming a commodity.

18.7.1.1 Rating Structures There are two structures used to determine the cost-per-minute rate that telcos charge to wireless carriers: banded rate structures or flat-rated rate structures. *Either method can apply to tariffs or contracts.*

The banded rate structure is based on an incremental increase in the cost to terminate mobile-to-land calls, by assigning specific rates to concentric mileage bands that emanate outward from a designated telephone company rate center (central office or tandem). In other words, if a wireless carrier orders a Type 1 interconnection that uses a banded rate structure, all mobile-to-land calls that are transported through that Type 1 interconnection are rated as "originating" out of that particular end office. Mobile-to-land calls whose termination point is within a certain "band" are charged at x cents per minute. A telco designates anywhere from three to eight concentric circular bands (circles) whose center is always the designated rate center. Each band encompasses a specific geographic mileage range outward from the rate center. Calls that terminate in mileage bands that are closer to the rate center are cheaper. Conversely, calls to destinations at the farthest mileage band away from the rate center are rated as the most expensive. Just 4 to 5 years ago, banded tariffs were used by many types of landline carriers, such as Ameritech, Southwestern Bell, and even interexchange carriers such as AT&T. However, because of intense competition in today's marketplace, more and more carriers—especially interexchange carriers—are using flat-rate pricing.

Flat-rate pricing reflects a very simple pricing mechanism: telcos charge carriers one flat rate (i.e., 3 cents per minute) to terminate calls across an entire LATA. It doesn't matter how far away from the interconnection point—the rate center—a mobile-to-land call terminates.

18.8   Special Construction Charges

When a wireless carrier orders any type of interconnection, special construction charges may apply. This usually occurs when the wireless carrier has a cell or MSC in a very remote or mountainous area. If the wireless carrier orders interconnection, and the telco cable facilities are a long distance away from the wireless carrier's location, then the telco will have to make an assessment. This assessment revolves around whether or not other entities (residences or businesses) will be able to make use of the newly required cable facilities within a given period of time (2 to 4 years). If no other entities will be able to make use of the cable, the cost to trench the cable must be borne by the wireless carrier. Depending on the amount of special construction that's required, the cost could be several thousand dollars, or even in the tens of thousands of dollars. If special construction charges are prohibitively high, the wireless carrier should seek other options for connectivity. These options could include seeking out other sites for interconnection, or more creative options such as working with the telco to implement a microwave radio link between the two networks.

18.9   Ordering Procedures for Interconnection to PSTN

In dealing with the RBOCs or large independent telephone companies such as GTE, all orders for interconnection are placed through special service centers known as *interexchange carrier service centers* (ICSCs). Usually the persons taking orders for interconnects are familiar with the language and concepts of interconnection. There are usually service representatives dedicated to certain wireless carriers, to allow for the development of a good working relationship between the two carriers.

18.10   Interconnection Agreements

There are two methods used by *landline* carriers to develop interconnection agreements with wireless carriers: either tariffs or contracts. Sometimes a combination of the two is used.

In most cases, facilities' rates are based on appropriate tariffs, and the per-minute-of-use charges are negotiated and may include tariffed elements. All interconnection agreements must be approved by state Public Utility Commissions (PUCs).

18.10.1   Tariffs

The cost-per-minute rates that the RBOCs charge to terminate (connect) calls to the PSTN are usually stated in the form of tariffs, which must be filed with and approved by the respective state PUCs. Tariffs state the rates charged for services, as well as the responsibilities of both the telephone company and the party that is purchasing service. Although it may seem odd that the party purchasing service has responsibilities, tariffs are designed in this manner to minimize the legal risk to telephone companies in case there are service problems.

When a wireless carrier purchases an interconnection from an RBOC, the tariffed rates apply for all charges. Tariffs state the cost-per-minute rates for carrying mobile-to-land calls, and rates charged for leasing circuits. There is no option to negotiate tariffs on a case-by-case basis when ordering interconnections from an RBOC. However, occasionally a consortium of wireless carriers that operate in a given state may band together in order to formulate a proposal for a new tariff structure for an RBOC's service area. The wireless carriers might designate one or more persons to meet with representatives of the RBOC to present their case for tariff restructuring. This situation usually occurs in cases where the rate structures are 2 to 3 years old. This time frame used to be longer, say 5 to 8 years, but with the explosion and growth of wireless technology, the shorter interval has become commonplace. This is because the intense growth in mobile traffic justifies lower rates for wireless carriers.

Once a proposal for tariff restructuring is presented to the RBOC, they review it internally. Once approved, the RBOC then submits the new tariff to the state PUC for approval. This process can take anywhere from 6 months to a year to complete. Tariffs are public documents.

18.10.2 Contracts

If interconnection rates charged to wireless carriers are not derived from a tariff, they are formalized through a written interconnection agreement: a *contract.* Contracts are usually used by independent telephone companies, such as Alltel or the Sprint telcos. With contracts, there is a large section at the beginning which gives a definition for all facets of the service offering. This is to preclude any charges of misrepresentation by the wireless carrier as to how and in what forms service is offered. Contracts are private documents.

18.11   Cost-per-Minute Rates

There are two methods that are used to determine the cost-per-minute rate that landline telephone companies charge to cellular carriers. Either method can apply to tariffs or contracts.

18.11.1   Banded Rates

Banded tariffs are used by all types of landline carriers. This rate structure is based on mileage bands emanating outward from a designated telephone company rate center. The rate center is the central office (geographic location) where the interconnect exists and is used for rating and measuring mobile-to-land wireless calls. In other words, if a wireless carrier orders a Type 1 interconnect in Idabel, Oklahoma, all mobile-to-land calls that are transported through that Type 1 interconnect are rated as originating out of the Idabel end office. For a one-time (nonrecurring) fee, the cellular carrier can change the rate center to another central office. This practice is done on the basis of traffic patterns.

Figure 18-8
Banded rate struc-
ture.

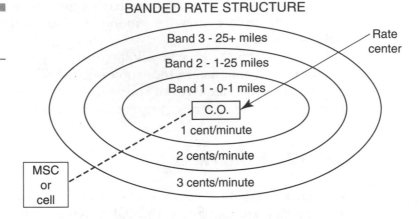

BANDED RATE STRUCTURE

The banded tariff structure works as follows. A landline carrier designates anywhere from three to eight circular bands (circles) whose center is always the designated end office/rate center. Each band encompasses a specific mileage range outward from the rate center. Mobile-to-land wireless calls whose termination point is within the band with the lowest amount of mileage from the central office are rated cheapest. Conversely, calls to destinations at the farthest mileage band away from the central office are rated the most expensive. As the mileage bands progress outward from the rate center, the cost to carry (terminate) traffic to destinations within that mileage band also increases correspondingly, on a sliding scale. For example, if a telco uses three mileage bands, the first band encompasses 0 to 8 mi from the rate center. This is known as "Band 1." The second band encompasses 8 to 25 mi from the rate center, or "Band 2." The third mileage band encompasses anything over 25 mi away from the rate center. Calls to someone living in Band 1 are rated the cheapest; calls to persons in Band 3 are rated the most expensive.

Banded rates are more difficult to administer and deal with for all parties involved. The more progressive telcos are implementing what is known as a *flat-rate tariff* structure. See Figure 18-8.

18.11.2 Flat Rates

Telephone companies such as U.S. West and BellSouth use flat-rate tariff structures when charging wireless carriers for calls made over their interconnections. Flat-rate cost-per-minute charges operate as follows. All mobile-to-land wireless calls are rated exactly the same from the rate center to any destination, LATA-wide. In other words, if a wireless carrier purchases a Type 1 interconnection from BellSouth in Madisonville, Kentucky, it pays the flat rate of 3 cents per minute to any destination in the LATA where the town of Madisonville is located.

There are several advantages to this type of structure. First, it is much easier to administer and deal with for both parties—the telco and the wireless carrier. Second, it can save the wireless carrier facility costs (the installation and monthly recurring charges for the interconnection circuit) by not having to obtain a Type 2 interconnection all the way to an access tandem, which could be considerable mileage. A third advantage that flat-rate tariffs provide is that, since they offer LATA-wide service, technically a wireless carrier doesn't have to install more than one interconnect in any given LATA. This would substantially decrease the monthly recurring costs for circuits and associated trunks over interconnects, by reducing the overall number of interconnects required.

18.12   Least-Cost Routing

Wireless carriers must ensure that all mobile-originated calls are routed to their destinations as inexpensively as possible. This process is known as *least-cost routing.*

The standard least-cost routing methodology usually employs *three* possible routes, cheapest to most expensive, for directing mobile-originated traffic from point A (the wireless network) to point Z (any land-line destination). If the first trunk group is busy (the cheapest option), calls will automatically be directed to the second trunk group (the second cheapest option), and then possibly to a third trunk group (the most expensive option) if the second group is also busy. This type of routing is accomplished by having a *route pattern* designated for every NPA-NXX combination in the country. The route pattern points to the three possible routes (trunk groups) that are available to send traffic. The route patterns exist in the switch—the MSC. (*Author's note:* All switches, whether PBXs, access tandems or toll tandems, have route patterns for all NPA-NXX combinations.) Obviously, the goal with least-cost routing is to ensure that the first trunk group is always properly engineered to handle most if not all calls, thereby avoiding overflow to the more expensive routes.

Key: The wireless carrier's own network should always be used whenever possible as the first route for mobile-originated traffic within wireless carrier *clusters.* (A "cluster" is a geographically contiguous group of wireless markets.) Ideally, mobile-originated wireless traffic should be directed over the wireless carrier's own network to the PSTN interconnect that is *closest to the destination of the wireless call.* For calls within a wireless carrier's own system, in a geographic cluster where the carrier has deployed its own signal transfer points (STPs), interswitch SS7/IS-41 links can and should be used for least-cost routing purposes. These links are known as *intermachine trunks* (IMTs), using SS7 ISUP (ISDN User Part) for call management.

18.13   The Current Evolution of the Public Switched Telephone Network

The nature of the PSTN is changing rapidly. In order to obtain cost-effective interconnection to the PSTN, familiarity with all current developments that are changing the face of the public telephone network is essential.

18.13.1   Number Portability

Number portability describes the ability of LEC customers to obtain service from a CLEC, while still retaining their original directory telephone number. This applies to residential customers, business customers, and wireless carriers as well (for numbers they obtained through interconnection to the PSTN).

Key: Regulatory deadlines for number portability to be active are as follows:

- By March 31, 1998, number portability capability was to be active in the largest metropolitan area of each Bell company region. For example, in the case of Ameritech that meant that number portability was to be active in the Chicagoland area by the date above.

- By December 31, 1998, number portability was to be active in the top 100 metropolitan areas nationwide.

The implementation of number portability goes hand in hand with the spread of local exchange competition. Competition will be stifled at a minimum, and nonexistent at best, without local number portability. If a wireless carrier wanted to obtain interconnection from a CLEC, the wireless carrier would need to have some means to be able to transparently move its (customer) telephone numbers over to the CLEC, if those numbers had been originally obtained from the incumbent telco (RBOC). Without that ability, true competition will not be realized, as envisioned by the Telecom Act of 1996. Local exchange competition and the existence of number portability will allow a wireless carrier to obtain service from (or convert to) a CLEC that will likely be offering lower rates for interconnection. This should ultimately reduce total costs that get passed on to wireless subscribers.

The flip side of the number portability situation is that eventually wireless subscribers may want to switch their service from one wireless carrier to another wireless carrier. Although the onus is currently on the landline telcos being able to cope with the issues surrounding number portability, it will inevitably be imperative for all wireless carriers to be able to migrate their subscribers' phone numbers over to other wireless carriers, with little effort.

Number portability will operate by having all carriers connect into a nationwide database via SS7 links. When callers dial a number that's been ported, SS7 messages will be exchanged between the two carriers through the national database. The call will then be routed to and terminated by the appropriate carrier—the carrier to which the number has been ported.

18.13.2   Interconnection Negotiations

Most wireless interconnection to the PSTN occurs within local exchange areas (intra-LATA). As the amount of competition to offer local exchange and interexchange services has increased dramatically, the ability to effectively negotiate interconnection rates and charges with any number of licensed carriers becomes critically important.

18.13.3   Cocarrier Status

Prior to passage of the Telecom Reform Act of 1996, one of the largest stumbling blocks that existed for wireless carriers in obtaining equitable and fair interconnection rates and terms was the fact that LECs did not grant wireless carriers "cocarrier" status in accordance with FCC orders. Cocarrier status dictates that when LECs determine interconnection rates and terms, they should treat wireless carriers just as they would treat another LEC or IXC in terms of how *those* carriers interconnect to the LEC's network and what rates *they're* charged per minute to transport calls into the LEC's network (i.e., RBOC to independent LEC interconnection, independent LEC to independent LEC interconnection, or IXC to LEC). Technically, it is illegal for any LEC to charge a wireless carrier a higher rate to terminate calls than it charges other carriers to terminate "like" calls into the LEC's network.

18.13.4 Reciprocal Compensation

Mutual compensation describes a concept that was pushed by wireless carriers, prior to passage of the Telecom Reform Act, to obtain agreement with landline telephone companies to pay the cellular company for *land-to-mobile* telephone calls. This issue heated up in the mid-1990s because the proliferation of wireless service resulted in a large increase in *land-to-mobile* telephone traffic. The wireless carriers came to the point where they felt it was only fair that they also be compensated for calls that terminate onto *their* network. When a wireless carrier transports a mobile-to-land telephone call, this call originates within the wireless network, and terminates to the PSTN (the landline network). Wireless carriers are required to pay a specified cost per minute for each of these calls. However, when landline customers called a mobile telephone number (land-to-mobile calls), the landline telephone company did not pay the wireless carrier any monies for terminating any of these calls onto the wireless network. Although this was a violation of FCC orders, the LECs got away with not paying mutual compensation to wireless carriers for years.

On December 15, 1995, the FCC issued a Notice of Proposed Rulemaking dictating that a concept known as *reciprocal* compensation should govern interconnection between commercial mobile radio service (CMRS) providers [cellular, PCS, paging, enhanced specialized mobile radio (ESMR) carriers] and local exchange carriers. Reciprocal compensation simply takes mutual compensation one step further, saying that under many (if not all) circumstances, landline and wireless carriers should bill each other for calls terminating to each other's networks. Reciprocal compensation is the ultimate maturing and resolution of the mutual compensation and cocarrier status issues.

The Telecom Act of 1996 formalized the reciprocal compensation issue by making it law. It stated that if a wireless carrier is being charged a certain cost-per-minute rate for calls terminating to the landline network, then the landline carriers have to pay the wireless carrier the exact same cost-per-minute rate for calls that terminate to the wireless network as well. For example, let's say that Verizon Communications (the former Bell Atlantic company) charges wireless carriers in New York City 1.2 cents per minute to terminate calls to the landline network. The wireless carrier can then charge Verizon 1.2 cents per minute for calls that originate from a Verizon landline phone and terminate to the wireless carrier's network.

An extension of the reciprocal compensation concept is known as "reciprocal termination," or "bill-and-keep." This concept is simply an extension of the reciprocal compensation concept. It describes a scenario where the amount of minutes of use (traffic load) between a given landline carrier and a given wireless carrier are about equal. Therefore, the carriers don't even bother to bill each other for the traffic that is carried between their networks. They bill their own customers for this traffic, and keep the monies without having to turn around and charge the other carrier. Hence the term "bill and keep."

Analysts agree that if costs are equal, the most efficient method of payment between two partners is no payment. There is almost no cost in the actual act of shutting down a line when one party hangs up the phone. Reciprocal compensation was the government's way of ensuring that the financial playing field between landline and wireless carriers was finally leveled off—fairness is ensured by the force of law.

18.13.5   Enhanced 911 Service (E-911)

Wireless customers account for 25% of all wireless 911 calls today (circa 2001). In October 1999, it was estimated that the 911 center in New Jersey handles 100,000 wireless 911 calls! To that end, it has become increasingly important for wireless subscribers to have a reliable means to be located by emergency service organizations when dialing 911 from a mobile phone.

In 1996, the FCC required a two-phase implementation of wireless E-911 services. It required wireless service providers to provide 911 service that is comparable to landline service providers. Phase 1 of wireless E-911, which was to be completed by March 31, 1998, required wireless carriers to provide 911 services to *any* 911 caller, subscriber, or nonsubscriber (valid or "invalid" mobile phones). Phase 1 of wireless E-911 requires the following to be offered by all wireless carriers:

- Routing of emergency calls to Public Safety Answering Points (PSAPs) based on the location of the mobile phone user.

- Delivery of automatic location information (ALI) and a callback number to the PSAP on call setup. ANI provides the number of the mobile phone to the emergency dispatcher (based on SS7 messaging).

- Access for speech- and hearing-impaired callers through the use of Text Telephone devices (TTY). Implementation of this requirement has been delayed due to wireless carriers' inability to provide appropriate devices.

The Phase 2 requirements of wireless E-911 are to be implemented by October 31, 2001. Phase 2 for *network-based* implementations (see Section 18.13.5.1, "Locator Technologies") include the ability to identify the latitude and longitude of a mobile phone within a radius of less than 125 meters 67% of the time and within 300 meters 95% of the time. The handset-based implementation requires 100% compliance of new or updated mobile phones by the end of 2004. The design and function of Phase 2 networks will build on the design of Phase 1 networks with the addition of geolocation capabilities. Geolocation capabilities are needed to provide latitude and longitude coordinates.

The FCC proposed the implementation of regional (state and local) cost mechanisms to cover the cost of implementing wireless E-911. The FCC also stipulates the legal liability of the wireless service providers—they have the same legal liabilities as the landline providers.

18.13.5.1 Locator Technologies There are three approaches wireless carriers can select from to determine the location of a mobile subscriber for E-911 purposes: network-based solutions, handset-based solutions, and hybrid solutions that utilize both approaches.

Network-based solutions employ triangulation and timing, using precise measurements of the wavelength differentials between base stations to reveal locations. One network-based solution uses time difference of arrival (TDOA) technology. With this approach, at least three base stations must be seen by the mobile handset to resolve ambiguities in location and they must be tightly synchronized. The base stations calculate the time it takes for the signals to travel from the stations to the handset, and back to the stations. This yields the distance of the handset from the base stations. The wireless systems will then map the three calculated distances, and wherever these locations meet reveals the position of the handset.

Another network-based technology being investigated is the direction of arrival (DOA) technique. DOA requires only two base stations, and the base stations must have an array of antennas. These base stations also calculate the distance of the handset from the base stations like TDOA. In addition, it also measures the *angle* of the signals being sent back to the base stations. The distances and angles measured together can then be used to calculate the location of the handset.

These network-based technologies are the most promising triangulation techniques and the easiest to implement; no change in standards is required; and, best of all, they work with unmodified phones. However, one drawback is that the mobile position significantly degrades in the urban environment due to signal reflection, and it loses accuracy when the mobile phone is close to base stations(s) and/or in hilly areas. Service providers are discovering that implementing these new services can be costly and time-consuming. Some estimates say the average cost to implement a network-based solution—per base station—is $500,000!

Handset-based solutions use the Global Positioning System (GPS) and an antenna on the handset to communicate user locations to the PSAP and the wireless service provider. This technology can pinpoint handsets within 15 ft outdoors and 100 ft inside buildings, which exceeds the FCC's requirements. GPS units are placed throughout a wireless network. As these units keep track of GPS satellites orbiting the Earth, they pass along key satellite information—including estimated time of the signal's arrival—to nearby wireless handsets, which are equipped with scaled-down GPS receivers. Then, based on time differences between when the network's GPS units and handsets receive signals from the satellites, it's possible to precisely pinpoint the caller's location. This option is too costly to implement, however, since GPS units must be included in the mobile handsets and these units consume a lot of power. The handset-based solution is more accurate and provides universal coverage. Yet the main drawback with this solution is that new handsets would have to be distributed to wireless customers. There is no easy way to implement this solution to the embedded base of customers.

The hybrid solution combines network-based and handset-based technologies.

Many industry experts tend to favor the handset-based solution because they believe that the network-based solution can be seen as an invasion of privacy. TDOA-type technology effectively turns the mobile into a de facto tracking device when it is powered on. With the network-based solution, the user can be located *anytime*.

18.13.6   Current Topics in Interconnection

There are several hot topics in today's interconnection world, circa 2000. First and foremost, the cost per minute charged to wireless carriers will continue to trend downward, to around less than 1 cent per minute, possibly as low as $\frac{1}{2}$ cent per minute.

Second, universal access to digital features when roaming (intelligent network development) will become a hot issue by 2002. This especially applies to emerging wireless technologies such as Wireless Application Protocol (WAP). The continued merging of wireless and data will bring a host of new issues to the table in terms of interconnection structure, rates, and Internet Protocol (IP) addressing.

Third, should the wireless industry be required to implement local number portability (LNP) in 2002, service portability could become a huge issue—especially if the wireless industry makes inroads into displacing wireline service in many American homes.

18.14   Test Questions

Multiple Choice

1. Type 1 interconnections to the PSTN:
   (a) Terminate directly to a telco access tandem
   (b) Terminate directly to a telco end office
   (c) Carry local/EAS traffic
   (d) Carry 911, 411, and operator-directed traffic
   (e) a and c only
   (f) b, c, and d only

2. The Telcordia Reference for Interconnection to the PSTN is:
   (a) TR-000-240
   (b) TR-140-000
   (c) Reciprocal Termination
   (d) GR-000-145

3. Which type of interconnection displays the ultimate case of least-cost routing?
   (a) Type 1
   (b) Type 2B
   (c) Type 2A
   (d) Type 2T

4. What is the difference between a banded interconnection structure and a flat-rated interconnection structure?
   (a) A banded rate structure charges one rate for calls terminating anywhere in a LATA; flat rate doesn't.
   (b) A flat-rated interconnection structure charges higher rates depending on how far away from the originating end office (or access tandem) the call terminates (connects). A banded rate structure charges different rates for different states.
   (c) There's no real difference between banded interconnection structures and flat-rated interconnection structures.
   (d) Banded interconnection structures charge higher rates, depending on how far away from the originating end office (or access tandem) the call terminates (connects). Flat-rated interconnection structures charge one rate, LATA-wide.

5. The reason that wireless customers are charged for land-to-mobile calls, even though they did not originate the call themselves, is because:
   (a) Wireless carriers are greedy, and have all engaged in price-fixing practices.
   (b) Land-to-mobile calls are still using extensive resources of the wireless carrier's infrastructure.
   (c) Wireless carriers are always charged for land-to-mobile calls also.
   (d) All of the above.
   (e) None of the above.

6. Dedicated interconnections to interexchange carriers are justified by doing a cost analysis to verify:
   (a) That there is always going to be 1 million minutes per month directed over the interconnection to the IXC.
   (b) That a Type 2A interconnection isn't necessary.
   (c) That the monthly cost for the DS1 circuit that allows for dedicated access to the IXC is equal to or less than the total cost of all mobile-originated calls that are routed to the IXC as "switched access" through a Type 1 interconnection to the LEC.
   (d) All of the above.

7. LATAs are designated by a three-digit identification. For LATAs that are dominated by independent telephone companies, that designation begins with the number:
   (a) 3
   (b) 6
   (c) 5
   (d) 1
   (e) None of the above

8. Wireless interconnections to the PSTN are also known as:
   (a) Localities
   (b) CLECs
   (c) POPs
   (d) IXCs

True or False?

9. _____ Telephone numbers are not purchased for sale to customers when a Type 2B interconnection is installed.

10. _____ Type 2A interconnections are direct links to telco end offices.

11. _____ Wireless carriers are required to purchase an entire NXX code when obtaining a Type 1 interconnection.

19

Cellular Call Processing

19.1 Enabling Technologies

In the late 1970s, several critical technologies came together simultaneously to propel the wireless industry forward. These technologies served as catalysts for the necessary technical advances in manufacturing to produce the first cellular telephones. Some of these technologies are:

- *Frequency synthesis function,* the ability of a mobile telephone to dynamically tune itself to any of the frequencies in its assigned spectrum, was shrunk down to the scale of a single integrated circuit (IC). This provided the means to do away with complex networks of crystal oscillators and tuning elements.

- *Microprocessors and microcomputers* (i.e., MSCs) allowed complex control functions such as cellular call processing and handling of user keypad functions to be performed within a very small space.

- *Surface mount technology* (SMT) applied to discrete circuit elements allowed for the shrinkage of linear circuitry blocks. This resulted in savings in weight and volume.

All of these factors came together to not only provide for small, relatively lightweight mobile phones, but to allow wireless switches to perform their functions as well.

19.2 Mobile Telephone Specifics

Mobile stations in AMPS systems are categorized according to their maximum allowed ERP (effective radiated power). Each mobile phone can be *commanded* to a power level within an operating range of levels by the base station; however, the mobile may not radiate more power than is allowed for its class. Power control levels are separated by 4 dB each.

Key: The capability of the mobile telephone to change power levels is known as *control channel mobile attenuation code* (CMAC).

CMAC describes the maximum power level that a mobile station may use when accessing the wireless system on the reverse control channel (mobile to base). It is set on a per-cell basis, and may be different from cell to cell.

The *station class mark* (SCM) of a mobile phone indicates the maximum ERP that the mobile can deliver, as well as the operating frequency range of the phone.

Each mobile telephone is identified by two unique numbers.

1. The mobile identification number (MIN), which is a 34-bit binary encoded representation of the mobile station's 10-digit directory telephone number. This number is programmed into the mobile phone when a subscriber obtains service.

2. The electronic serial number (ESN), which is a 32-bit binary number that is programmed into the mobile phone in the factory at the time of manufacturing. This 32-bit number can yield over 4 billion combinations. The ESN is unalterable, and uniquely identifies the mobile telephone. The ESN contains an 18-bit serial number and an 8-bit manufacturer code, and is viewed as the electronic fingerprint of the mobile telephone. The remaining 6 bits are reserved for future use.

19.3   Mobile-Originated Calls

Once the mobile phone has been powered on in an AMPS system, it monitors the 21 possible control channel frequencies in its assigned band (A or B band) prior to the placement of a call. The purpose of the monitoring is to find the control channel with the strongest signal and lock onto that signal to watch for overhead information and paging information. In a properly engineered wireless system, the strongest signal will come from the nearest cell site.

1. When the user originates a mobile call, the mobile telephone will send in a data message on the reverse control channel to the closest cell site, with the dialed digits entered, the MIN of the phone, and the ESN. This activity will only occur when the busy/idle status (BIS) indicates that the reverse control channel is idle. The base station reverse control channel then receives the origination message and forwards the message (via the cell controller) to the MSC.

2. The MSC receives the origination message, verifies the MIN/ESN combination as valid, and initiates seizure procedures on a PSTN interconnection trunk, assuming the destination of the mobile-originated call is a landline telephone. The MSC then allocates a traffic channel at the cell site where the user is located that will bear the call. The traffic channel is the voice channel (VCH) that carries the actual conversation. The MSC then sends a VCH assignment message to the cell base station with the identity of the VCH.

3. Upon receiving the VCH assignment message, the cell base station VCH keys its transmitter and begins sending supervisory audio tone (SAT) to the mobile phone. Simultaneously, the forward control channel sends an assignment order to the mobile station containing the frequency and the SAT of the VCH that will bear the call, and the MIN that identifies the mobile telephone.

Key: Maintenance of link continuity between the base station and the mobile phone in a traffic channel (VCH) is accomplished via the SAT tone, which is added to the voice signal prior to the modulation of the RF carrier.

4. The mobile telephone receives the assignment order and retunes its frequency synthesizer; then it monitors for the designated SAT.

5. Upon SAT confirmation, the mobile station keys its transmitter and regenerates the SAT to the base station.

6. The cell base station detects the regeneration of the SAT by the mobile station and sends an *origination complete* message to the MSC. The MSC then connects the PSTN trunk to the VCH trunk. The conversation may now take place, since both the cell base station and the mobile station have reached the same state.

Key: Since this is a mobile-originated call, when the MSC is connecting the VCH trunk (voice channel) to the PSTN trunk, least-cost routing has already occurred. The MSC has examined the dialed NPA-NXX digit combination, and determined the least expensive route to direct the call.

See Figure 19-1 for a step-by-step diagram of mobile-originated call processing.

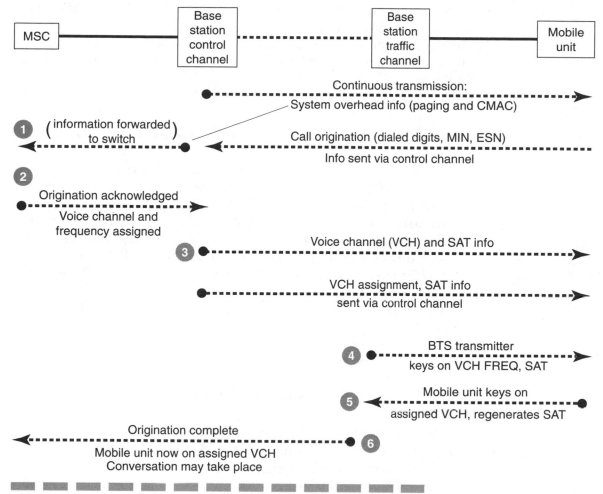

Figure 19-1 Mobile-originated wireless calls: transactional signaling process.

19.4 Mobile-Terminated Calls

19.4.1 Overview

Wireless customers are charged for land-to-mobile calls (calls they did *not* originate themselves) because the land-to-mobile call still uses considerable resources of the wireless network infrastructure. Land-to-mobile calls use interconnection trunks to access the wireless network as well as cell radio channels, to fully process a land-to-mobile call. To that end, wireless carriers are justified in charging their customers to terminate these calls. This is even easier to understand and justify considering the industry average cost for a wireless base station is anywhere from $250,000 (PCS) to around ($500,000 to $750,000) (cellular).

19.4.2   Autonomous Mobile Registration

Autonomous mobile registration plays an integral part in the processing of mobile-terminated wireless calls. Also known simply as *mobile registration,* or *registration,* it is the process where, by design, all wireless phones continually transmit their MIN/ESN combination over the air to the nearest cell site every 5 to 15 minutes so that the MSC will know the mobile user's location in the event of an incoming call. Mobile registration is the mobile phone's way of continuously saying "Here I am" to the wireless system.

There are more steps involved in the processing of land-to-mobile calls when compared to the processing of mobile-to-land calls. This is mainly because of the paging function and the alerting function.

19.4.3   Call Processing (Mobile-Terminated)

Similar to a mobile-originated call, the mobile station continuously monitors the 21 control channel frequencies in its assigned band (A or B band) and locks onto the strongest control channel.

Scenario: A person sitting at home decides to call a friend and they initiate a call over the PSTN by dialing the 10-digit number (MIN) assigned to the mobile station. The PSTN routes the call to the appropriate interconnection associated with the MIN, and on to the MSC.

1. The MSC checks its database for a list of the last 10 cell sites where the mobile station last registered and broadcasts a page to all those base stations. The base station where the mobile is located receives the page message and transmits it to the mobile phone on the forward control channel. The mobile telephone receives the page message while monitoring the control channel.

2. The base station forwards the page response to the MSC. The MSC receives the page response message, verifies the MIN/ESN combination, and allocates a VCH (voice channel) at the cell site that will bear the call.

3. The MSC then sends a VCH assignment message to the base station with the identity of the VCH (i.e., channel set/frequency assignment).

4. Upon receiving the VCH assignment message, the base station VCH keys its transmitter and begins sending the SAT to the mobile phone. At the same time, the control channel sends an assignment order to the mobile telephone that contains the frequency and SAT of the VCH that will bear the call.

5. The mobile telephone receives the assignment order and retunes its frequency synthesizer, and monitors for the designated SAT. Upon SAT confirmation, the mobile station keys its transmitter and regenerates the SAT to the cell base station.

6. The cell base station detects the regeneration of the SAT by the mobile station and sends an alert order to the mobile telephone, which causes the mobile phone to ring.

7. The mobile station confirms the alert status by continuously gating the signaling tone into the reverse VCH signal.

8. At this time, and after detecting the presence of the signaling tone from the mobile station, the cell base station sends a message to the MSC indicating that the mobile station has successfully arrived on the VCH and is alerting.

9. When the wireless customer goes off hook, the mobile telephone removes the signaling tone from the reverse VCH signal.

10. The base station detects the absence of the signaling tone and sends a message to the MSC indicating that the mobile telephone has gone off hook. Conversation may now take place, since both the base station and mobile telephone have reached the same state.

See Figure 19-2 for a step-by-step diagram of mobile-terminated call processing.

19.5   Call Handoff

Assume a wireless customer has placed a call from his phone to a friend across town, and he is in the middle of a conversation.

Scenario: The cell where the user is currently located, base station 1, detects that the reverse voice channel signal strength justifies the consideration of a handoff. This is determined via the received signal strength indicator (RSSI) at the cell base station.

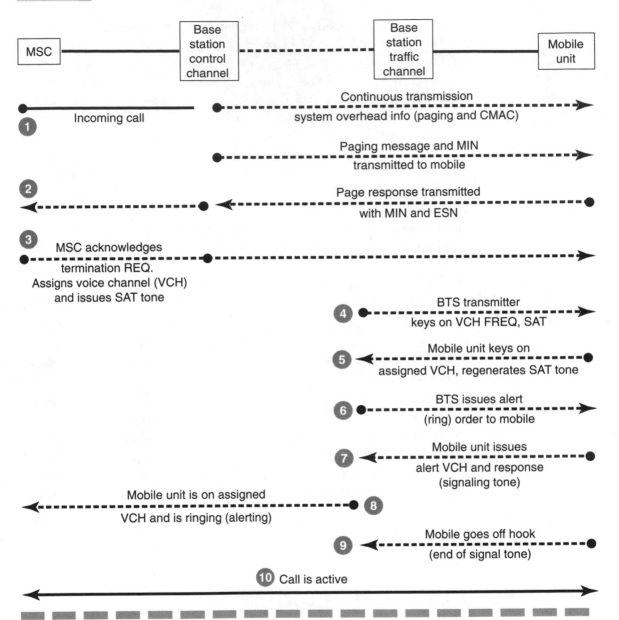

Figure 19-2 Mobile-terminated wireless calls: transactional signaling process (applies to land-to-mobile and mobile-to-mobile calls).

Key: The RSSI threshold parameter that determines when a handoff is justified is manually set by wireless technicians. However, it is at a threshold that is well above a power level that would allow a call-in-progress to be dropped from the system.

1. Base station 1 then sends a handoff request to the MSC containing information relevant to the mobile telephone (i.e., power class, current power level) and its *current* reverse voice channel signal strength.

2. The MSC receives the handoff request and determines which cells are adjacent to base station 1. In this example, base station 2 is one of several adjacent cells, and it is the one cell receiving the strongest signal from the mobile. Therefore, the MSC sends a handoff measurement request to base station 2. Base station 2 receives the handoff measurement request and uses its locating receiver to determine the suitability of a handoff. The locating receiver will tune to the reverse VCH frequency the mobile telephone is currently using, and make signal strength measurements on all (or some) antennas.

3. If the measurements exceed a certain criteria for handoff, then base station 2 will send a handoff measurement response to the MSC. The MSC receives the handoff measurement response and chooses a VCH for base station 2 to use to support the call.

4. The MSC then sends the VCH assignment message to base station 2. Base station 2 receives the VCH assignment message, keys the VCH transmitter (transceiver), and sends the VCH assignment confirmation message to the MSC.

5. The MSC receives the VCH assignment confirmation message from base station 2 and sends the handoff order message to base station 1 containing the frequency and SAT of the VCH in base station 2. Base station 1 transmits the handoff order to the mobile telephone over the forward VCH.

6. The mobile telephone hears the handoff order, confirms the order by "gating" 50 ms of signaling tone into the reverse VCH signal, and reprograms its frequency synthesizer for the VCH in base station 2. The mobile telephone then looks for base station 2's VCH SAT and, having found the SAT, regenerates it on the reverse VCH to base station 2.

7. Base station 2 detects the regeneration of the SAT on the reverse VCH and sends a *handoff OK* message to the MSC.

8. The MSC then sends a *release source channel* message to base station 1, which causes the forward VCH to be dekeyed, making the channel available for use by another subscriber in base station 1's territory. The call handoff is now complete, and the conversation is now being carried over a VCH at base station 2.

See Figure 19-3 for a step-by-step diagram of call handoff processing.

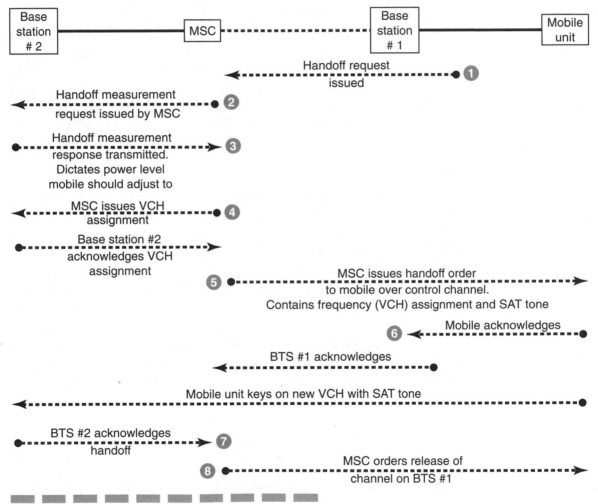

Figure 19-3 *Call handoff: transactional signaling process.*

19.6   Test Questions

Multiple Choice

1. The purpose of supervisory audio tone (SAT) is:
   (a) To dictate acceptable power levels to the mobile phone
   (b) To establish and maintain link continuity between the cell base station and the mobile phone
   (c) To initiate seizure of a PSTN trunk for the termination of mobile-originated calls
   (d) To inform the system (cell and MSC) when a call handoff is required

2. Which term or acronym below is used to indicate the maximum ERP allowable for mobile phones?
   (a) SAT
   (b) Transceiver
   (c) Station class mark (SCM)
   (d) CMAC
   (e) Filter
   (f) Duplexer

True or False?

3. _____ The RSSI threshold parameter that determines when a call handoff is justified is manually set by carrier technicians.

4. _____ Electronic serial numbers are embedded into the memory of mobile phones at the point of sale, at carrier business offices.

5. _____ SAT is added to the voice signal prior to the modulation of the RF carrier.

6. _____ The capability of the mobile phone to change power levels is known as CMAC.

20

Roaming and Intercarrier Networking

20.1 Roaming Overview

Roaming is the use of wireless service outside the subscriber's "home" service area. In other words, wireless customers are in a roaming state when they use their mobile phone outside of the area where they originally purchased their wireless service. Each wireless carrier determines its own home service area on the basis of the specific market where they are operating. In each market, one carrier may try to maintain a *larger* home area footprint (coverage area) in order to gain a competitive edge over its in-market competitor. The home service area may be designated on a large wall map in business offices where service is initially obtained by wireless subscribers. Once the mobile phone is powered on, customers become aware they are in a roaming service area through the autonomous mobile registration process. When the mobile phone is powered on in a roaming service area, the phone registers with the system where it is currently located. A *roaming indicator* will then appear on the liquid-crystal display (LCD) screen of the mobile phone, in the form of the letter "R" or the word "roam."

20.2 Rates and Charges

Special rates apply to wireless customers when they make roaming calls. These rates are very carrier-specific, and may fluctuate greatly between carriers across the United States. The rates are very high compared to regular call rates, but all roaming rate elements will continue to come down as more advanced intercarrier networking systems are implemented nationwide (e.g., ANSI-41, SS7). The only mitigating factor concerning the high charges that apply for calls placed when roaming is that *all* carriers impose high fees for roamer calls. Circa 2001, there has been a rapid development in the wireless industry regarding roaming charges, along with long distance charges. In order to gain competitive advantage, many carriers have adopted a "no roaming fees, no long distance charges" strategy. If this marketing strategy continues to take hold, roaming fees as they are known today will be a thing of the past by 2004.

But a typical rate structure for roamer calls is as follows, for carriers who still apply these charges:

- A *daily* fee is assessed, whether a person who's roaming makes one call or 20 calls per day. This fee can be as high as $3.00 per day.

- A *cost-per-minute* fee of anywhere from 50 cents to $1.50 applies.
- A *per-call* fee may be assessed. This fee could be 50 cents to $1.00. This fee is rarely imposed, however.

20.3   Old Roaming Structure

From 1983 to the mid-1990s, someone who wanted to call a person who was roaming would have to follow a cumbersome multistep process in order to reach the person. This process was so customer-unfriendly that roaming activity was minimal. The process was actually a *deterrent* to increasing roaming revenues.

1. A caller who wanted to reach a roaming customer would have to know the exact destinations of the person who would be roaming.
2. The caller would have to dial a special roamer access telephone number which would place them *into the switch* (MSC) of the market the cellular customer was roaming in. This number was obtained through a bulky paper directory issued by cellular carriers.
3. Then the caller had to dial the roamer's 10-digit mobile phone number, once they accessed the "roaming" MSC.

This roaming structure still exists today in some small markets, although it is becoming very rare because of the introduction of more advanced intercarrier networking systems. These systems enable callers to reach roaming customers *anywhere*, transparently, by simply dialing their 10-digit mobile phone number. These advanced systems are enabled via ANSI-41 and SS7 signaling systems.

20.4   Present-Day Roaming Systems

Although the majority of roamer traffic is to markets that are adjacent to a customer's home service area, this has changed with the increasing popularity of the portable wireless phone and the advent of more advanced roaming arrangements using networking technologies such as ANSI-41 and SS7.

20.4.1   Home Location Register

The home location register (HLR) is a database that is resident within every wireless switch. Its purpose is to acknowledge and maintain subscriber information relating to the identity of valid customers in the switch's *home* area. This occurs when home subscribers power on their phones and register with the system via autonomous mobile registration (over the control channel). The HLR will maintain information on every active (powered-on) mobile in its system. A "valid" customer is one who pays bills on time and whose MIN/ESN has not been declared illegitimate because of fraud (see Chapter 21, "Wireless Fraud").

Wireless carriers also have the option of implementing what is known as a "centralized" HLR. This is one HLR—one database—for the carrier's entire network. If a wireless carrier owns many wireless markets, the MSCs from all these markets can all be connected (over the carrier's own WAN backbone) to the one HLR. In these cases, the HLR will usually be at a geographic location that is central to all its market holdings. The advantage to having a centralized HLR is that it is easier and more efficient to maintain one HLR, rather than a multitude of HLRs. There will also be more consistency with how the HLR is programmed and maintained if there is only one of these machines. This scenario would also free up the MSC to handle what the MSC is meant to handle: the processing of wireless calls. Therefore, the MSC will also operate more efficiently. The only possible drawback to having one centralized HLR is that a carrier would have to closely monitor the throughput on its internal WAN backbone to ensure that the HLR (validation) traffic is not overwhelming the capacity of the WAN. In a worst-case scenario, the wireless carrier would have to expand its WAN backbone, which could be a costly undertaking.

20.4.2   Visitor Location Register

The visitor location register (VLR) is a database that is also resident within every wireless switch. Its purpose is to acknowledge and maintain subscriber information relating to the identity of valid *roaming* customers in a switch's home area.

Once a mobile phone is powered up in a nonhome (roaming) system, it will immediately perform a mobile registration. Once the phone registers onto the system, the switch obtains subscriber information from the customer's home switch (via an SS7 network), and all information pertinent to that roaming subscriber is stored in the VLR (e.g., feature sets).

20.4.3   ANSI-41 Signaling Systems

Seamless networks enable wireless subscribers to place calls from virtually any place and maintain uninterrupted wireless service—whether subscribers move about in a geographic region served by multiple wireless operators or travel from one region (or market) to another. Prior to the mid-1990s, seamless roaming capabilities were available only by relying on switching equipment from a single manufacturer because the equipment communicated in a proprietary messaging format. In other words, if cellular carriers wanted to build internetworking roaming capabilities between themselves to support "seamless" roaming, both carriers had to use wireless switches from the same vendor.

The thrust for multivendor, seamless networks led to the development of an industry standard for intersystem networking that is known today as ANSI-41D. The ANSI-41D standard has evolved through several revisions since the early 1990s, as an interim standard (IS-41 Revisions 0, A, B, and C). It was formally standardized as ANSI-41D in late 1997.

IS-41 Revision 0 defined the methods for handoff interactions and precall validation between separate and distinct wireless systems serving different regions. To complete intersystem handoff when neighboring cellular markets are involved, a point-to-point DS1 link was required between the MSCs of the two providers. (Precall validation is a means to thwart wireless fraud. See Chapter 21, "Wireless Fraud.")

IS-41 Revision A (IS-41A) utilized HLRs and VLRs to house and manage the transfer of subscriber profiles, providing the base to standardize the methods for automatic roaming, automatic call delivery, and intersystem handoff. This included the beginning of the subscribers' ability to use the same system features that they're normally capable of using in their home systems while they're roaming.

IS-41 Revision B (IS-41B) defined additional enhancements for roaming, including Global Title Translations (GTT), handoffs for TDMA digital systems, and new messages, which support the use of three-way calling and call waiting for calls that have undergone an intersystem handoff.

IS-41 Revision C (IS-41C) offered the latest enhancements in intersystem signaling. It built on IS-41B by standardizing services expected to be most commonly used by subscribers while roaming, such as short message service, authentication, and calling number identification presentation and restriction. Capability for automatic call delivery in border cells for AMPS and TDMA digital systems is also included in IS-41C. This revision of the IS-41 standard can support today's cellular systems as well as PCS systems with a great deal of flexibility.

ANSI-41D incorporates all of IS-41C functionality into a formal standard and is the latest evolution of the industry standard. It was published in December 1997. ANSI-41D defines support for authentication, short message service, automatic call delivery in border cells, message waiting notification, and caller identification (ID), to name a few. Most wireless carriers today provide SS7 messaging capabilities to connect with other wireless carriers across the country and the world, to allow ANSI-41D messages to be sent over SS7 TCAP (Transaction Capabilities Application Part) facilities, which provides a reliable, high-capacity message transport medium to support intercarrier networking.

Key: The ANSI-41 standard for interswitch signaling is *not* restricted to links between *different* wireless markets. In a given market where there is more than one MSC in the market, these MSCs can also be connected via ANSI-41 signaling over the carrier's own internal network. The purpose of this setup is to transport intrasystem calls from one MSC to another MSC seamlessly. This use of ANSI-41 signaling enables a carrier to avoid using the PSTN for intrasystem call transport, thereby reducing their interconnection costs. This is a fundamental example of least-cost routing.

20.4.4   Signaling System 7

Signaling System 7 (SS7), known internationally as Common Channel Signaling System 7, has been in use in the wireline networks since the late 1970s and early 1980s. SS7 is a data signaling network (overlay) that is separate and distinct from the wireless carrier's radio and interconnection trunks that are used to carry subscriber traffic. There are multiple functions of the SS7 protocol in networks today. First, a key purpose is to optimize call routing and processing in such a way as to reduce the total amount of facilities necessary to process calls. Second, SS7 also serves to ensure that least-cost routing is always a factor in mobile-originated call routing. Third, SS7 allows for the deployment of enhanced services such as caller ID, automatic callback, call waiting, and other "vertical" services offered by all types of telephone companies, both wireline and wireless.

Throughout the 1980s, wireless carriers had their MSCs connected together in a meshed architecture to achieve roaming capabilities with each other. As wireless network growth exploded, a hubbed architecture became necessary to increase efficiencies and cut costs. Wireless carriers have been implementing SS7 on a large scale since the early 1990s, to allow for nationwide roaming for wireless customers. As of 1997, wireless carriers across the United States had implemented about 50 mated pairs of STPs to facilitate seamless roaming for their customers.

In today's wireless networks, SS7 infrastructures are used to transport ANSI-41D messaging to other wireless carriers to support the key features that ANSI-41D allows for, namely (automatic) call delivery and intersystem handoff. The ANSI-41D messages are *encapsulated* in the SS7 messages and transported across a nationwide SS7 backbone network to support seamless roaming [see Section 20.4.6, "(Automatic) Call Delivery"].

20.4.5 Signal Transfer Points

Special packet data switches known as signal transfer points (STPs) are necessary elements of SS7 network infrastructures. In reality, STPs have a dual functionality: they act as both hubs and packet switches. To allow for SS7 functionality in the wireless world, carrier's MSCs are connected to STPs. Physically, STPs can be built-in adjuncts to the MSCs, or they can be stand-alone systems. This configuration depends on the maker of the mobile switch (MSC) as well as the engineering practices of the wireless carrier. Larger wireless carriers procure their own STPs, and then have all their MSCs connected to their STPs. These links are implemented using 56-kbps or DS1 circuits and are known as "A" (access) links. This is how the STP acts as a hub. Smaller, regional wireless carriers sometimes lease STP space from the larger carriers. See Figure 20-1 for a depiction of "A" link functionality.

Just as the wireline telephone companies deploy their STPs in "mated pairs" in order to achieve equipment redundancy, wireless carriers also deploy their STPs in mated pairs. All telephone companies need the redundancy that is afforded by having a backup STP. The STPs must be connected to each other, and these links can be implemented by using either 56-kbps data circuits, multiples of 56-kbps circuits, or full DS1 circuits. The link that attaches the two members of a mated pair of STPs to each other is known as a "C" (cross) link. See Figure 20-2 for a depiction of "C" link functionality.

To attain a national footprint that allows for termination of calls to mobile subscribers who are roaming, wireless carriers then connect their own STPs to national SS7 providers' STPs in order to obtain (nationwide) access to other wireless carriers' networks for the purpose of connecting

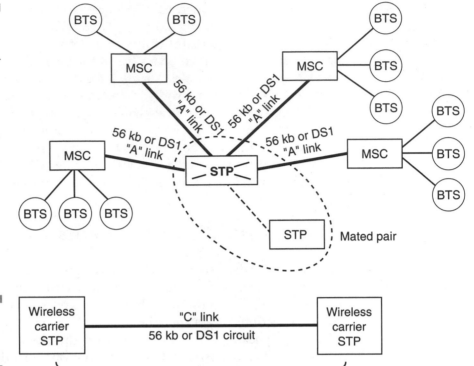

calls to mobile roamers almost anywhere in the United States. These connections from a cellular carrier's STPs to a national SS7 provider's STPs are known as *"B" (bridging) links*. They are usually implemented using 56-kbps data circuits. A 56-kbps data link from one mated pair of STPs to another mated pair of STPs is known as a *B-link quad* because four STPs are actually all connected together, in a meshed configuration. This configuration allows for full redundancy of the STP pairs between the wireless carrier and the nationwide SS7 signaling providers. See Figure 20-3 for an illustration of a B-link quad between a wireless carrier mated STP pair and a nationwide SS7 provider's mated STP pair. It should be noted here that the linking of STP pairs and the associated transport of traffic between carriers follows a hierarchical architecture. "B" links are called "bridging" links because they are building a bridge from one type of carrier, or a regional wireless carrier STP pair, to a carrier that has a nationwide network footprint. When carriers connect their own STP pairs together—for redundancy purposes—those links are known as "D" links. This includes any two carriers, whether or not they are wireless carriers.

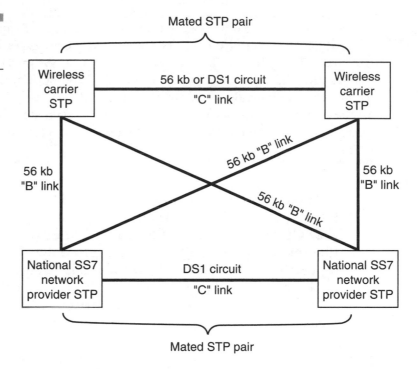

Figure 20-3
"B" links and B-link quad.

It should also be noted that all of these links in the SS7 world, including the (redundant) "A" links that connect switches to STPs, have redundancy built into their operations as well as their architectures. Each network element has two connections into other network elements. All of these links are also load balanced, meaning that the amount of total traffic is split roughly 50-50 onto each link. Therefore, each link operates at only 40% capacity. That way, if one of the links goes down for some reason, the other link can sustain the traffic generated by *both* links, with 20% excess capacity for traffic spikes.

Examples of national SS7 network providers are North American Cellular Network (NACN) and Illuminet [formerly Independent Telecommunications Network (ITN)]. NACN mostly services direct links from A-side carriers, while Illuminet mostly services direct links from B-side cellular carriers. GTE also provides nationwide SS7 connectivity and transport services. Any of the providers also serve PCS carriers.

In order for one wireless carrier to send calls (for its roamer customers) from its network to another carrier's wireless network across the country, both carriers must be connected into a national SS7 provider's network at a minimum of one location.

Key: If a cellular carrier owns both A markets and B markets, it can route all SS7 connectivity from its own mated pair of STPs into just one national SS7 provider's STPs if it chooses. That one SS7 provider can then *gateway* calls destined for mobile customers of the opposite band carrier to the other nationwide SS7 network provider if necessary. In some form, all the nationwide SS7 network providers have their mated STP pairs connected into the other nationwide SS7 network providers STP pairs in D-link quads.

Example: A cellular carrier owns both A-side markets and B-side markets nationwide. This cellular carrier has a mated pair of STPs at its Network Operation Center (NOC) location. The carrier connects its STPs to Illuminet (serving mainly B-side carriers) only. Illuminet has its STPs connected to NACN STPs. Through special arrangements, the cellular carrier can ensure that traffic destined for A-side mobile customers (who are ultimately served by the NACN STPs) will terminate safely to the intended wireless network.

Key: All wireless carriers' STPs are identified by unique addresses known as *signaling point codes* or *point codes* for short. Point codes are equivalent to the IP addressing of the SS7 world. Any device that handles SS7 traffic must have a point code assigned to it.

20.4.6 (Automatic) Call Delivery

Automatic call delivery is the process of seamlessly transporting a call to a roaming customer who is in another wireless market. The market where the customer is roaming could be anywhere in the United States where coverage exists and the carrier has SS7 connections into a nationwide SS7 provider. The call is transported *seamlessly* because callers do not need to know the location of the mobile customer they are calling. Callers also don't need to dial a special *roamer access* telephone number.

The automatic call delivery process works as follows:

1. When a person roams into another wireless market and powers up the mobile phone, it immediately performs an autonomous mobile registration with the "visiting" wireless system. As part of the registration process, John Doe's system identifier (SID) is downloaded to the VLR in the roaming switch. The SID identifies the person as a subscriber of another carrier's system, and includes a special identification for that person's home carrier. The roamer MSC then sends an ANSI-41 message (encapsulated in an SS7 packet) to John Doe's home MSC to inquire if he's a valid customer. The home MSC then informs the roaming MSC that John Doe is a legitimate customer via HLR records, and forwards a copy of his feature set to the roamer MSC. This signaling transaction is a form of handshaking and validation.

2. At that point, the roaming system's MSC knows that John Doe with MIN (*XXX-XXX-XXXX*) and ESN (*YYYY*) is now resident in its system. If someone from John Doe's home market attempts to call him by dialing his mobile phone number, his home MSC will send an ANSI-41 message over the nationwide SS7 network, notifying the visiting MSC that there's an incoming call for John Doe. The roamer MSC then assigns John Doe a temporary listed directory number (TLDN). The TLDN comes from a pool (a range) of numbers that is set aside from a linc range that the roaming system's carrier obtained via interconnection in their market.

3. The roaming system then signals back to John Doe's home system (using SS7 signaling), informing the home system's MSC that John Doe has a temporary number in its system, and the number is *ZZZ-ZZZ-ZZZZ*.

4. At that point, sincc John Doe's home MSC knows he's in the visiting system with a TLDN of *ZZZ-ZZZ-ZZZZ*, the home MSC routes the call to the roaming system using a combination of the PSTN and the nationwide SS7 network. The home MSC routes the call to the roamer system via its dedicated IXC trunks, while an SS7 message is sent to notify the roaming system that the call is on the way.

5. Once the call reaches the roaming MSC, the paging process begins, and the call is routed directly to John Doe's mobile phone (to the TLDN number), as if he were a home subscriber in that market.

The TLDN's purpose is to act as a pointer to John Doe's actual home system MIN. It acts as an alias phone number while John is roaming in that "foreign" wireless network. That way, calls to John's wireless number are pointed directly to the roaming carrier's MSC via the TLDN, and are eventually terminated to John's phone in the market where he is roaming. This is all made possible because John Doe's home MSC has *all* nationwide NPA-NXX combinations in its routing tables. Once the roamer MSC informs John Doe's home MSC what the TLDN is, the home MSC knows

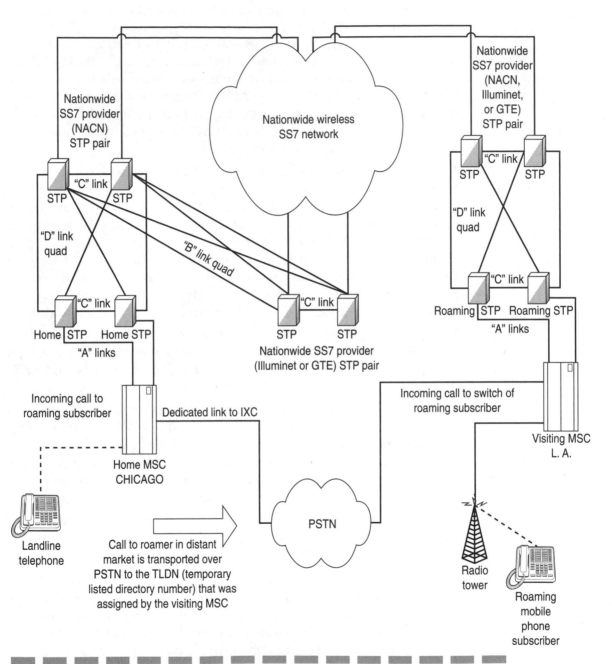

Figure 20-4 Automatic call delivery and nationwide SS7 network architecture and operation.

to route the call over its long distance trunks because of the (least-cost routing) directions in its routing tables. See Figure 20-4 for a depiction of automatic call delivery and the nationwide SS7 network.

Key: If someone attempts to call John Doe, that person does *not* need to know that John is roaming or even where he is located.

20.4.7   Intersystem Handoff

Intersystem handoff (IHO) is made available through ANSI-41 signaling. Intersystem handoff is the process of transparently handing off a call in progress from one wireless system to a neighboring wireless system. This process is made possible because of an MSC-to-MSC link between the neighboring markets. This link is known as an intermachine trunk (IMT).

The intersystem handoff process is very similar to the regular (intra-market) call handoff process, and works as follows:

1. Engineers from wireless carriers whose markets are next to each other must select certain cells on their market borders which will be used for handing off calls between the markets.

2. A point-to-point DS1 circuit must then be installed between the MSCs in the two markets. Intersystem handoff requires a minimum of a DS1 circuit.

3. If John Doe is on the phone in his home system and he is traveling toward the neighboring wireless system, the cell that's carrying his call sends a message to the home MSC that John Doe's signal is getting weaker. This is determined via RSSI, just as it is with intrasystem handoff.

4. One of the cells that's identified as a handoff candidate will be a border cell site in the *neighboring* system. This will be a cell that was selected for handoff by engineers from both companies.

5. John Doe's home MSC will then send a handoff request to the neighboring system's MSC over the point-to-point (DS1) link between the two MSCs.

6. The two switches will exchange handoff signaling parameters, and the call will be handed off to the neighboring system over the DS1 connecting the two MSCs. In that instant, the call is transferred from the home system cell to a border cell in the neighboring carrier's system.

7. The call in progress has been handed off from one system to the neighboring system transparently.

Key: The call is carried over the point-to-point DS1 between the two markets, and stays up on that DS1 until the call is terminated.

20.5   International Roaming

With the world shrinking at the rate it is in lieu of the globalization of the world economy and the blazing pace of evolution in the information age, international roaming for wireless carriers has started to become a major action item for forward-looking wireless carriers. GSM carriers have a much easier time of it than carriers using any other digital wireless technology, because GSM—by default—requires strict adherence to its standards. This, along with the fact that there are more GSM carriers worldwide than any other type of wireless carrier, makes international roaming in the GSM world almost effortless. It is very easy for the GSM carriers across the world to link their systems together. GSM roaming revenues are running at a rate of around $550 million per month!

International roaming agreements are not too difficult to reach per se. But once roaming agreements are signed between two carriers from different countries, the technical aspects of the partnership take longer to work out. Tests and trials must be run with each partner. This can take about 4 to 6 weeks for *each partner.* One of the most trying aspects of the testing process is running trials with partners that operate in vastly different time zones. During these trials, carriers can run about 20 call scenarios to determine whether various features will function properly while roaming with the partner. It is also very important for international roaming partners to deliver call detail record (CDR) files to each other in a timely manner to ensure accurate and rapid billing.

One example of a very successful international roaming endeavor is BellSouth Mobility's venture with 36 independent service providers throughout Latin America. A key fact that rationalized this push was that there are 350 airline flights per day in and out of Miami alone that originate or terminate in Latin American cities.

To maximize the opportunities to increase revenues through international roaming, wireless carriers need to find partners with a style and pace that suit their customers' international roaming needs. But other options exist. Service providers can choose to join an alliance, for example (i.e., the GSM Alliance).

20.6 Wireless Intelligent Network (WIN)

The *wireless intelligent network* (WIN) is a concept being developed by the Telecommunications Industry Association (TIA) Standard Committee. The charter of this committee is to drive intelligent network (IN) capabilities, based on the ANSI-41D standard currently embraced by wireless providers because it facilitates roaming. Basing WIN standards on the ANSI-41 protocol enables a graceful evolution to an IN without making current network infrastructure obsolete.

All the technologies already discussed in this chapter represent types of WIN applications. IN is required for various (precall) validations and billing reciprocation of wireless calls. Along with transparent roaming services, selective call screening and acceptance; short message service; speed-to-text conversion; and prepaid billing are just some of the examples of the types of features and functionality that a WIN infrastructure can provide.

20.6.1 WIN Functions

The WIN mirrors the wireline intelligent network (IN). The distinction between the wireless and the wireline network is that many of the wireless call activities are associated with *movement*, not just the actual phone call. In the WIN, more call-associated pieces of information are communicated between the MSC and a device known as a service control point (SCP), or the HLR. The WIN moves service control away from the MSC and up to a higher element in the network, usually the SCP.

In a WIN, the SCP provides a centralized element in the network that controls service delivery to subscribers. High-level services can be moved away from the MSC and controlled at this higher level in the network. It is cost-effective because the MSC becomes more efficient, does not waste cycles processing new services, and simplifies service development.

In a WIN, there is also a device known as an intelligent peripheral (IP). This device gets information directly from the subscriber, be it credit-card information, a personal identification number (PIN), or voice-activated information. The IP gets information, translates it to data, and hands it off to another element in the network—like the SCP—for analysis and control.

The following steps will need to occur before WIN will be a reality:

- Incorporation of SCPs and IPs into the wireless network architecture.
- Evolution of the MSC to an SSP. In the IN, the SSP is the switching function portion of the network. The mobile switching center (MSC) provides this function in the WIN.
- Separation of call control and transport from service control.
- Development of generic call models, events, and trigger points.

20.7   Test Questions

Multiple Choice

1. The links that attach the two members of one mated STP pair are known as:
   (a) C links
   (b) A links
   (c) B links
   (d) D links
   (e) F links
   (f) None of the above

2. Which interim standard describes interswitch networking?
   (a) IS-54
   (b) IS-95
   (c) IS-45
   (d) ANSI-41
   (e) None of the above

True or False?

3. _____ Automatic call delivery (ACD) involves the transparent routing of a call to a mobile telephone from a cellular customer's home market to a market where the customer is roaming.

4. _____ Use of the ANSI-41 standard for interswitch signaling is restricted to links between different cellular carriers.

5. _____ "D" links describe the links from a cellular carrier's MSCs to its own STP (or STPs).

21

Wireless Fraud

21.1   Overview

According to the Cellular Telecom Industry Association (CTIA), in 1997 the wireless industry lost $434 million—$1.18 million daily—to wireless fraud. These losses for 1997 represent 1.4% of total industry revenues. In 1998 losses were estimated to be $182 million, or 0.05% of industry revenues. This equates to a 1-year drop of 58% in projected losses to the industry. There are several reasons for the decline in losses. First, the cellular industry's migration to digital wireless technologies has proceeded at a very rapid pace. Digital radio systems are inherently more fraud resistant, mainly due to the inability (thus far) to clone digital radio handset transmissions. Second, the implementation and use of authentication systems to fight fraud has risen dramatically since the mid-1990s. Third, most wireless carriers now operate fraud-profiling systems in their networks. These systems fight fraud on several fronts, through the use of velocity checks, ANSI-41D signaling systems that are used in conjunction with validation of roamers, and other measures.

Also, the rapid attainment of market share by the PCS providers since 1996 has ultimately decreased the market share of cellular carriers. Since PCS systems are all-digital and employ authentication systems from the start, fraud in PCS systems has been less of a problem than it's been for cellular carriers.

A history of the different types of fraud, and the countermeasures that the industry employed to deal with them, will be presented.

21.2   Tumbling ESN Fraud

"Tumbling ESN" fraud was the first type of cellular fraud. The original cellular switches in the United States were not networked together by the ANSI-41 (or SS7) protocol. They had no means to communicate with one another. To employ tumbling ESN fraud, "fraudsters" programmed multiple random mobile identification numbers (MINs) from different geographic areas into a phone, and made up a false ESN using a memory chip [an electronically programmable read-only memory (EPROM) chip]. At that time, if a switch received a service request from a phone not based in that area (a roamer), it would allow the first call to be placed and go through. Simultaneously, the "roamer" switch would send an inquiry message to a nationwide clearinghouse and check the nationwide "negative" database—a database of illegal wireless phone numbers—to see if the phone was stolen or voided because of nonpayment. This scenario reflects a system where cellular carriers had dedicated data links (usually 56-kb circuits) to the nationwide clearinghouse. The clearinghouse would charge the carriers a monthly fee for this "validation" service. Companies such as GTE and EDS operate some of the major clearinghouses. If the phone's MIN/ESN combination was not resident in the negative database, the roamer switch would allow additional calls to be placed. If the phone was registered in the database, no additional calls could be placed by the roaming customer. The fraudster would get the first call off, and then "tumble" the ESN, looking like a brand new caller to the roaming system's switch (VLR). Once the new ESN registered with the system, the fraudster looked like a brand new customer and could place another call, and so on.

The negative database at the nationwide clearinghouse was manually updated by the carriers. Some carriers were very diligent about updating the records in the database, while others were not so diligent. Eventually, the systems would automatically update every night at 12:01 A.M.

In the early days of cellular, criminals sold tumbling ESN phones on the streets of New York for as much as $300. Cellular carriers effectively attacked this type of fraud by requiring "precall validation" in their switches, the MSCs.

21.3   Precall Validation

Most carriers today use some form of precall validation, where both the MIN and ESN of the mobile phone have to be verified through national services (operated by GTE) before a call is processed by a roamer switch. Most of this verification activity today is handled through SS7 networks. The validation can be processed through a nationwide clearinghouse (database), or wireless carriers can process the validation directly between themselves (MSC to MSC) using the nationwide SS7 backbone networks now utilized by most wireless carriers (see Chapter 20, "Roaming and Intercarrier Networking"). It is presumed that when carriers used the nationwide clearinghouse method, many carriers would be connected into the clearinghouse to maximize its effectiveness. This obviously included most, if not all, of any given carrier's roaming partners. Only phones with properly "validated" MIN/ESN combinations are allowed to place calls. With precall validation, a roamer switch (the particular switch that a call origination request is directed to) will initiate a query into the service bureau's database to see if the MIN/ESN pair is valid, *or* send a request to the home switch to validate the legitimacy of the person (mobile) attempting to place the call. These procedures have all but eliminated instances of tumbling ESN fraud.

Ultimately, it became too cumbersome for the carriers to rely on the clearinghouse databases. They weren't consistent with each other. GTE had most of the B-side carriers, and EDS had most of the A-side carriers. The roamer information had to be shared via a gateway, and many times the gateway was down or not functioning properly. Because of this, a company named SystemsLink invented a product which allowed different mobile switches to communicate with each other and eliminate the clearinghouse altogether. This product became known as *Roamex*. Roamex handles most of the A-side carriers' roaming records and information database, as well as most of the B-side carriers'. Roamex's only competition is a product developed by GTE. SystemsLink handles 98% of the cellular market to date, and all of Global System for Mobile Communication (GSM) North America. Roamex allows for near real-time record exchange which means the carriers not only can validate customers to determine their validity, but they can also see what calls are being placed within minutes, sometimes seconds. This feature allows profiling systems to be much more accurate when detecting fraud (see Section 21.5.1, "Profiling Systems").

21.4 Cloning Fraud

Cloning is the complete duplication of a legitimate mobile identification, namely, the MIN/ESN pair.

Around 1991, criminals began scanning the airwaves and stealing MIN/ESN pairs from the air. With a simple scanner set to the right frequency (824 to 894 MHz), and a setup near a busy location such as an airport or interstate overpass, an enterprising crook could collect hundreds of valid MIN/ESN pairs quickly.

From a historical perspective, cloning has mainly been a problem for analog (AMPS) systems in the cellular world, where it is easy to intercept transmissions. Instances of cloning fraud have decreased since the mid-1990s, due to cellular carrier's migration to digital radio technologies and widespread implementation of clone detection techniques.

The first step in cloning fraud is to obtain a complete mobile identity. Unfortunately, this information can be easily obtained as it is transmitted on the control channel every time a mobile phone registers (autonomous mobile registration), originates a call, or receives a call. Then, with a computer and a few cables, the electronic components that store the ESN can be altered and programmed into the crook's phone.

Since the ESN itself is unalterable once it is programmed on a chip, what cloning criminals do is remove the chip and install a blank EPROM chip. A stolen ESN and MIN pair is then programmed into the blank EPROM chip, yielding a completely cloned, working cellular phone. A cloned phone has had the MIN, ESN, and, in systems that use it, the PIN completely replicated.

The PIN is a 4-digit personal identification number. Use of PINs was being employed extensively in the mid-1990s in either of two ways to combat cloning fraud:

1. Mobile subscribers dial a feature access code (e.g., star 56) and then their 4-digit PIN just to enable themselves to make a call.

2. Mobile subscribers then enter their PIN immediately *after* they have completed a call to resecure their phone's ability to place calls.

Of the two ways that PINs are used, method 1 is much more cumbersome and annoying. Once the call is made, the customer must reactivate a different feature access code, along with the PIN, to relock the phone so no calls can be made.

Key: Call processing using pins prevents unauthorized calls from being made at the switch level within the MSC itself.

Cloned phones are produced by crooks and sold, usually in distant markets that are not home to the cloned phone, to other crooked and/or unsuspecting persons. (This is known as "clone and roam" fraud.) Because the switch will check to ensure MIN/ESN pairs are valid, this cloned phone appears valid to the unsuspecting switch (usually a roamer switch) until the appropriate cellular carrier or legitimate customer determines that a valid phone has been cloned.

Mobile customers can tell if their phones have been cloned when they receive calls from strangers. This happens because the crooks using the cloned MIN/ESN unwisely distribute their MIN (cellular phone number) to their associates. Then, their associates attempt to call them. But they may reach the valid customer instead, if the cloned phone happens to be turned off, and the legitimate one is turned on, since there are two active phones with the exact same MIN/ESN pair. Obviously, the point here is that crooks should use a cloned phone only to originate calls, and not to receive calls. Another way that mobile customers can tell if their phones have been cloned is if their service is suspended. When attempting to place a call, the customer receives a recorded announcement stating that he or she should contact a billing representative immediately. When doing so, the billing representative informs the customer that the phone has been cloned, and that crooks have racked up charges of X number of dollars. The carrier has suspended the phone service because their profiling system has determined that the usage on the (cloned) phone was abnormal for some reason. In these cases—when a customer's phone has been cloned—the industry response to the customer is similar to the approach of credit card companies. The customer is only held responsible for a maximum of $50 of the total fraudulent charges.

In the early to mid-1990s, over 40,000 cellular subscribers a month got their phones cloned. Many times, wireless carriers are also subjected to cloning fraud due to "dumpster diving" by criminals. In these cases, crooks go through the trash to try to find MIN/ESN information about legitimate cellular subscribers. Internal fraud by dishonest salespeople, or simply clever con artists, also ends up costing wireless carriers significant revenues. Sometimes, computer hackers can even access a carrier's billing system and steal MIN/ESN pairs.

Today, persons who clone cellular phones are labeled counterfeiters. Because of this label, the Secret Service has worked with the CTIA and cellular carriers to fight cloning fraud.

For many years cloning was the costliest form of fraud. However, as previously stated, because of the widespread deployment of digital cellular and all-digital PCS systems, and the implementation of authentication in many wireless switches, today cloning is one of the fastest-shrinking forms of wireless fraud.

21.5   Industry Response to Cloning Fraud: Clone Detection Techniques

There are several techniques by which cloned phones can be detected. Some of the major techniques are described below, along with their impact on cellular customers. There are many vendors that provide clone detection systems today.

21.5.1   Profiling Systems

In response to cloning, cellular carriers can put profiling systems in place that would set off system (database) alarms when suspicious MINs break certain thresholds or show abnormal behavior compared to their *historical* recorded use. Examples of certain thresholds or abnormal behavior could be a very high amount of calls within a short period of time, a large amount of calls of long duration, a large amount of "roamer" calls, or a large amount of international calls.

One of the most common forms of profiling is known as a *velocity check*. Systems that use the velocity check method measure the time between two call attempts or mobile registrations with a given MIN/ESN, which took place at two distant geographic locations. If the two registrations took place just minutes apart, and the physical distance between the two registrations is, say, 600 miles, then the profiling system would issue a "flag," telling persons monitoring the system that this mobile phone has most likely been cloned. Velocity checks confirm the physical impossibility of two mobile registrations taking place hundreds of miles apart within minutes of each other.

Profiling allows carriers to shut off a cloned phone fairly quickly by suspending its ability to place calls. Profiling systems must be purchased by wireless carriers, and integrated into the MSC. Persons must be hired and trained to staff the profiling system. Profiling systems don't automatically shut off suspicious users. The persons monitoring the system must make educated assessments to determine whether cloning fraud has occurred.

21.5.1.1   Problems with Profiling   The main problem with the profiling method of detecting cloned phones is that it only works *after* calls have been completed, because of its reliance on postcall validation. An owner of a cloned phone could make many calls before the MIN/ESN has been suspended by the cellular carrier. All those calls equate to lost revenue. A legitimate customer would still receive a bill that had calls made by the owner of the cloned phone. The legitimate customer would then need to bring his or her phone into a cellular service center and have a new MIN assigned to it. Although most cellular carriers absorb the majority of the cost of the illegitimate calls that were made from a cloned phone, it may take a day or more to detect cloning fraud, even with a profiling system.

Although profiling does not stop fraudsters completely, they become frustrated with profiling because their cloned phones work only temporarily, and they have to continue to clone more phones. This impacted the cellular industry in a positive way because, although it had only minimal impact on the instance of cloning fraud, it reduced the *amount* of illegitimate calls made by the owner of a cloned phone. The time frame where cloned phones are functional has been compressed, and the amount of illegitimate minutes of use generated on the network goes down.

21.5.2   RF Fingerprinting

The concept of RF fingerprinting began when three companies with a history of U.S. defense radio research and development claimed that they could recognize most phones by an *individual pattern* of radio transmission. One of the main companies providing such a service is Corsair, Inc. These claims appear to be true, since RF fingerprinting has a high success rate at recognizing fraudulent mobiles (80 to 90%).

RF fingerprinting began widespread deployment around 1995, and today at least 50% of the top 100 MSAs in the United States are using RF fingerprinting to detect (and eliminate) home-market fraud. Cellular One uses RF fingerprinting in its Chicago MSA, and U.S. Cellular uses RF fingerprinting in some of its Texas markets, especially those markets near the Mexican border.

RF fingerprinting operates by sampling and digitizing the individual patterns of radio transmissions. These samples are then stored in the local MSC. When a call is attempted, the RF of the mobile phone attempting to originate a call is sampled and matched against the characteristics of the mobile that are stored in the database of the local MSC, along with the MIN/ESN pair. If no match occurs between the RF fingerprint and the MIN/ESN pair, no calls are allowed to be processed.

21.5.2.1   Problems with RF Fingerprinting   There are several drawbacks to this method of detecting cloned phones:

- Customer service is impacted because of false positives (1 to 2%). This problem is being examined by the industry and could be solved by a network that transmits a digitized RF transmission from a home system to a visited system (over ANSI-41 links).

- It is possible that cloners may find certain models of phones where this technique does not reliably work.

- Difficulties occur with roaming because the three fingerprinting techniques being used are incompatible, and not every market may be able to afford RF fingerprinting. Until RF fingerprinting becomes a ubiquitous technique, its ability to detect cloned phones won't be optimized.

21.6   Authentication

Authentication is the process where a switch verifies the identity of a cellular phone before processing a call. The authentication antifraud process has been touted as the ultimate way to thwart cloning fraud.

21.6.1   The Authentication Process

Authentication refers to a challenge/response process during which information is exchanged between the mobile phone and the switch (MSC) for the purpose of confirming the mobile phone's identity. The authentication process is governed by the Cellular Authentication and Voice Encryption (CAVE) algorithm. The mobile phone inputs several parameters into the algorithm and sends the output to the switch. The switch then makes the same calculation and compares the two results. If the results match, the phone (the phone's identity) is believed to be authentic. When a user turns on a wireless phone, the phone uses the authentication "key" and other parameters to calculate the shared secret data (SSD). This shared secret data travels over the airwaves (and the fixed network) to the switch for verification. The switch also knows the authentication key for that subscriber and does the same calculation that the mobile handset did, and compares the data to ensure the results match. If they match, the switch authenticates the phone and processes the call. Only the switch and the handset know the authentication key, also known as the "A" key.

Sending a password over the airwaves does not happen when using the authentication feature. Authentication is a superior method to protect against fraud because it does not rely upon the open transmission of the ESN to verify subscribers for call billing information.

Authentication *services* are also offered by certain companies, where mobile subscribers who are roaming are intercepted by the service when they attempt to originate a call. The service directs the caller into an interactive voice response (IVR) system and asks for information (already in the IVR's associated database) that only the caller would know (i.e., social security number, mother's maiden name). Once the caller's response matches the information in the IVR database, the caller may proceed with placing the call in a roaming environment. Sometimes roaming wireless customers are also forced to use credit cards as a means to thwart fraud. These can be phone credit cards or Visa/Mastercard.

21.6.2   Industry Commitment to Authentication

Authentication is now an industry standard, firmly supported by carriers and phone manufacturers alike. Since fraud is an industry-wide problem, for authentication to work, it will require an industry-wide commitment. Although the issue of building to the new standard is a voluntary option for mobile phone manufacturers, most say they plan to support it and the industry's efforts to fight fraud regardless of the cost. In fact, the industry is so firmly supportive of the new authentication standard that the CTIA and phone manufacturers are urging the FCC to require that all new equipment in the future be built with the authentication feature (called *standardized encryption*). Furthermore, the CTIA may make adherence to the authentication standard a prerequisite for gaining CTIA certification.

Motorola, which actively participated in the authentication standards development process, made significant manufacturing modifications that were necessary and phased in authentication across *all* product lines in the fourth quarter of 1995, beginning with its digital mobile phones.

Small cellular carriers may be in an unfavorable position regarding implementation of the authentication feature. This is because a system reconfiguration to process authentication could be cost-prohibitive. Estimates place the reconfiguration cost per carrier, depending on its market size and number of switches, at $1 million plus. For smaller market carriers not yet feeling the pinch of fraud, implementation could be a long way off.

There are several incentives for even the smallest of carriers to implement authentication, although it will be costly. For one, the authentication algorithm also provides voice privacy, affording carriers a strong marketing hook. Providing fraud-free service will also give cellular carriers a leg up on local market PCS competitors, who will have a built-in advantage at providing fraud-free service from the start. In the meantime, the cellular industry has a long and expensive road ahead in dealing with fraud and deploying authentication. There are more than 28 million existing cellular subscribers who are currently unprotected.

21.6.3   Problems with Authentication

Although some carriers believe that authentication can provide the solution to cellular fraud that they have been seeking for years, there are problems with implementing the authentication method:

- There are millions of analog mobile phones that are currently in use today that do *not* have the authentication feature built in. Therefore, fraudsters will concentrate their efforts on cloning phones that do not have the authentication feature.

- Cellular carriers that do not have switches capable of authentication will not be able to use the authentication feature. Therefore, fraudsters will be able to steal numbers from one city and use the numbers in another city where the switches do not have authentication capabilities. Because this could impact roaming revenues, it would also eventually impact roaming agreements themselves that exist between carriers.

For authentication to be completely successful in eliminating cloning in the future, all carriers must reconfigure their switches to allow for processing of the authentication algorithm, and all mobile phone manufacturers must build to the standard. Carriers who don't or can't implement the authentication feature risk being frozen out of the (wireless) marketplace and alienated by their roaming partners.

Once it becomes too technically difficult to compromise the wireless network, fraudsters will take a different approach. They will try to get on the systems as a normal subscriber would, but using false identification. Hence, the fraud process would go full circle from the most technical type of fraud (cloning) to the least technical type of fraud: subscriber fraud.

21.7   Subscriber Fraud

Every year in the United States, there are 400,000 cases of "identity theft." This occurs when people's social security numbers and/or other personal information is stolen and used to commit fraud of one type or another. The wireless industry is not immune to this type of fraud, as fraudsters will use false identities to obtain wireless service.

Most subscriber fraud is "true identity" fraud, where the fraudster uses the stolen identity of a real person to obtain accounts with wireless carriers. In this case, the fraudsters use some type of false identification. These persons have no intention whatsoever of paying their bill when they initially apply for service with a wireless carrier. This type of fraud has increased dramatically since it has become more difficult to compromise wireless networks in a technical manner because of the deployment of authentication and digital wireless technologies.

Even with credit checks and official identification cards (driver's licenses, social security cards, etc.), people intending to commit subscriber fraud are able to get by the system. Many times people who commit subscriber fraud move from place to place to avoid detection.

Subscriber fraud is now becoming the largest fraud loss to date, occurring in both analog AMPS systems as well as PCS systems.

21.7.1   Industry Response to Subscriber Fraud

There are several software packages available today to minimize subscription fraud. These packages offer real-time validation of the identity of potential wireless customers as they apply for service. The systems allow the wireless carrier to interface with a third-party company's database which lists persons (criminals) who have a history of committing subscriber fraud in the industry. These third-party software systems will also interface directly with a carrier's customer activation system. If any of the information looks suspicious, the system can notify the sales agent to get more information or request a deposit before signing people up for service. Another option would be to offer these people prepaid service.

Preactivation systems check such information as driver's license mismatches or bogus addresses (e.g., mail drops, prisons); social security numbers issued before birth; social security numbers that have a death claim against them; and zip codes that do not match the city name.

Many prepay programs have also resulted in significant subscriber fraud losses. Prepay accounts have been responsible for up to 30% of fraud losses to some wireless carriers. This could be carried out by any of several means, as follows:

1. Illegally using prepaid card numbers prior to their purchase (e.g., physical theft, pattern checking, or computer hacking of the prepaid system).

2. Upon legal redemption of a prepaid card, the rightful owner's phone becomes cloned.

21.8   Test Questions

Multiple Choice

1. How much did the United States cellular industry lose *daily* to fraud in 1997?
   (a) $250 million
   (b) $5 million
   (c) $1.18 million
   (d) $700 thousand
   (e) $1.3 billion

2. Which type of fraud is defined as the complete duplication of a legitimate mobile identification?
   (a) Subscriber fraud
   (b) Cloning fraud
   (c) Tumbling ESN fraud
   (d) Magic phone fraud

3. Profiling systems are sometimes used as a fraud countermeasure. These systems also may use what is known as a *velocity check.* What does a velocity check do?
   (a) Measures the geographic distance between two locations where one mobile identification (MIN/ESN) has registered within a short amount of time, determining the physical impossibility of such activity
   (b) Recognizes certain mobile phones by the individual pattern of radio transmission

    (c) Determines if a person may be using a stolen social security number, false ID, etc.

    (d) None of the above

4. Authentication:

    (a) Is the process whereby a switch verifies the identity of a cellular phone before processing a call

    (b) Is sometimes referred to as the "silver bullet" that will eliminate cloning fraud

    (c) Refers to a challenge/response process

    (d) All of the above

22

Digital Wireless
Technologies

22.1   Overview

As demand for viable, commercial mobile telephone service has increased dramatically since its inception in 1983, service providers found that basic engineering assumptions borrowed from wireline (landline) networks did not hold true in mobile systems. This became evident due to the fact that approximately 28,000 people per day signed up for cellular service from the early to late 1990s. This number has increased dramatically since then. So in the mid-1990s, the critical problem for cellular carriers became *capacity.*

Cellular providers had to find a way to derive more capacity from the existing multiple-access methods currently in use, namely frequency division multiple access (FDMA). A multiple-access method defines how the radio spectrum is divided into channels and how channels are allocated to the many users of the system. Multiple-access technology allows a large number of users to share a common pool of radio channels, and any user can gain access to any channel.

The spectacular growth in the number of wireless customers has to be accommodated through a continual increase in system capacity. The most extreme and costly method is to *reduce* cell sizes (coverage areas) and introduce additional base stations—in effect, implementing cell splits. However, in most large cities (MSAs), it has become increasingly difficult and costly to obtain the necessary permits to erect base stations and antennas. Therefore, cellular carriers wanted a solution that made it possible to increase system capacity significantly without requiring more base stations. The solution was to introduce digital radio technology, which allows for increased spectral efficiency without an increase in the total number of base stations. In March 1988, the Telecom Industry Association (TIA) set up a subcommittee to produce a digital cellular standard. By the early 1990s, IS-54 systems were deployed using time division multiple access (TDMA) technology.

The implementation and use of digital wireless systems result in better use of available radio spectrum and cleaner, quieter signals than the AMPS analog systems. Digital transmission also provides for greater security against eavesdropping and cloning fraud.

There are three major digital radio technologies developed and in use today: TDMA, GSM, and code division multiple access (CDMA). Figure 22-1 depicts the three types of digital wireless systems, how they compare functionally, and how they utilize spectrum.

■■■ ■■ ■■ ■■

Figure 22-1
Multiple-access
methods for wireless
technologies. FDMA:
Users are assigned
different frequencies
within a given band.
This is an access
technique for analog
systems. TDMA:
Users are assigned
different frequencies
and different time
slots. CDMA: Users
are channelized by
specified codes with-
in a frequency band
that is 1.25 MHz
wide.

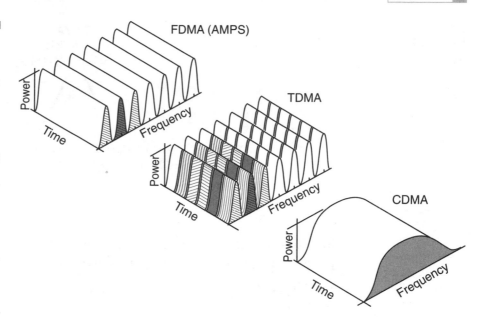

22.2   Digital Wireless Systems Comparison with Analog Wireless Systems

The two key advantages of digital cellular technologies over analog cellular networks include increased capacity and security.

There are some basic differences between digital radio systems and
analog radio systems:

1. Given equal amounts of spectrum, all of the digital wireless technologies allow for increased use of radio spectrum, compared to analog systems. The basic intention of deploying digital radio technologies is to
get maximum use of allotted radio spectrum. To that end, all of the digital radio technologies (TDMA, GSM, and CDMA) allow for use of a substantially larger amount of radio channels than analog AMPS systems.

2. Digital radio base stations cost significantly less than analog base stations. This is because digital base stations employ advanced, smaller-scale
equipment that has a smaller footprint as well as a smaller price tag.

3. All-digital systems can employ handoff techniques in which the
mobile handset has more of a role in determining if and when a handoff is required, and to what *cell* or *cells*. This improved handoff capability

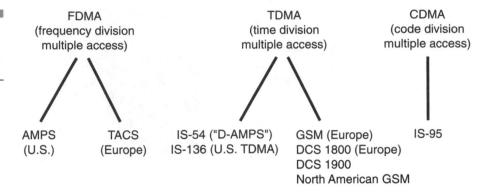

Figure 22-2
Digital wireless tech-
nologies and associ-
ated classifications.

increases the efficiency of the wireless network because it decreases the potential that calls will be dropped during the handoff process.

See Figure 22-2 for a classification of analog and digital wireless technologies.

22.3 Vocoders

The deployment of digital cellular systems has been enabled through the development and implementation of equipment known as *vocoders.*

Human speech is easy to reproduce. Vocoders are equipment (chip sets) embedded in mobile handsets and in base station transceivers that digitize human speech. Vocoders are what actually enable digital wireless systems to exist; they are the foundation that allows for increased use of radio spectrum. Vocoders sample transmissions of human speech, packetize the samples, and send digital pulses containing the packetized samples from the handset to the base station instead of always having the mobile's transmitter "key" on the appropriate frequencies. Distant-end vocoders (at the base station) decode the pulses and restore the speech to its original form.

In GSM systems, a full-rate vocoder allows for eight users (conversations) on a single 30-kHz radio channel. Hence, there is an 8:1 increase in capacity compared to analog AMPS systems.

In TDMA systems (IS-136), a full-rate vocoder allows for three users (conversations) on a single 30-kHz radio channel. Hence, there is a 3:1 increase in capacity compared to analog AMPS systems.

Key: Half-rate vocoders are able to sample human speech at half the rate of standard, full-rate vocoders. Therefore, there is a 2:1 ratio in increased utilization compared to full-rate vocoders. Half-rate vocoders sample human speech with less bits per sample than full-rate vocoders, thereby doubling their capacity. In GSM systems, a half-rate vocoder allows for 16 conversations (users) per channel. In TDMA systems, a half-rate vocoder allows for six conversations (users) per channel. However, it is important to note that half-rate vocoders produce higher bit error rates, and when not accompanied by more sampling, produce a substandard voice quality.

22.4   Time Division Multiple Access

22.4.1   Overview

Time division multiple access is the generic name for an air interface technology that is used by a variety of standardized digital radio systems, such as IS-136 and GSM. TDMA was the earliest incarnation of digital radio technology. It assigns both different frequencies and different time slots to each conversation on a wireless system. GSM is a digital wireless standard in its own right, even though it uses TDMA as its *air interface* technology. This is because GSM has many special features and attributes which make it a distinct radio technology.

IS-54 was the earliest standard to use TDMA technology, defining the migration path from analog to digital radio systems. The IS-54 standard, also known as the D-AMPS (digital AMPS) standard, referred to TDMA and digital radio in very generic terms and contained few references to digital radio feature sets. It focused on the migration from AMPS analog systems to digital cellular systems.

The revised, updated standard for TDMA systems, IS-136, contains information on full-featured TDMA digital systems, with references to feature sets such as caller ID and short message service (SMS), which provides text messaging.

Currently, IS-136 provides three voice calls per 30 kHz of bandwidth. Compared to the analog AMPS standard, this is a 3:1 increase in system capacity per channel.

TDMA systems can be more susceptible to interference than other radio technologies when used in combined AMPS/TDMA systems by cellular carriers, where calls could break up or mute. However, major improvements in vocoders since the early 1990s have allowed for vast improvements in dual-mode TDMA/AMPS transmissions.

For the most part, TDMA as a technology is being supplanted today by CDMA because CDMA offers many benefits over TDMA, and is ultimately more cost-effective because more capacity can be gleaned by an equivalent amount of network infrastructure. The one carrier today that continues to use and grow TDMA in its network is AT&T Wireless. This is due to the fact that when AT&T purchased McCaw cellular in the mid-1990s, McCaw had already deployed TDMA in most of its network. So AT&T made a business decision to continue the use of TDMA since the majority of the purchased embedded base—the former McCaw network—was TDMA. It is generally believed that CDMA, or a combination of CDMA and TDMA technologies, will become the de facto standard for future wireless networks (i.e., 3G).

22.5   Code Division Multiple Access

22.5.1   Overview

Code division multiple access (CDMA) is an American digital standard that was developed by a company named Qualcomm, based in California. CDMA was originally deployed as a battlefield communications system because it is very hard if not completely impossible to intercept CDMA transmissions. The CDMA digital scheme is titled Interim Standard 95 (IS-95). Many PCS carriers selected CDMA as their official digital wireless standard, such as Sprint PCS and Primeco PCS.

22.5.2   How CDMA Works

CDMA is a wideband, spread spectrum technology. A unique code is assigned to all speech bits (conversations). Signals for all calls are spread across a broad frequency spectrum, hence the term "spread spectrum." The dispersed signals are pulled out of what appears as background noise by a receiver that knows the code for the call it must handle. This technique allows numerous phone calls to be simultaneously transmitted on one radio frequency. As a result, CDMA systems can handle between 10 and 20 times the calling capacity of conventional cellular systems.

There is a certain analogy that is frequently used to explain how CDMA works. Imagine you are at a reception being held in a room at the United Nations. There are four people around you, each speaking a different native language: Spanish, Korean, Chinese, and English. Your native language is English. You only understand the words of the English speaker, and tune out the Spanish, Korean, and Chinese speakers. You hear only what you know and recognize. The same is true for CDMA. Multiple users share a frequency band at the same time, yet users only hear their own conversations.

When a mobile phone call is made using CDMA technology, the sound of the user's voice is converted into a digital code. This digital signal is first "correlated" with a noiselike code known as a pseudorandom noise (PN) code, also called "Walsh" codes. The correlator yields an encrypted digital representation of the original signal. This encrypted signal is then spread over a very wide frequency spectrum (1.25 MHz). At the receiving terminal, the signal is "demodulated" back to a narrow bandwidth, and then fed into a "decorrelator." This decorrelator uses its unique PN code to extract only the information intended for it. A signal correlated with a given PN code and decorrelated with the same PN code returns the original signal. Decorrelating the signal with the wrong PN code would result in pure noise, containing no discernible information or sound.

22.5.3 CDMA Architecture and Operations

In CDMA systems, the standard is to use 1.25-MHz blocks of radio spectrum to carry many conversations, using pseudorandom noise codes. Because of the nature of spread spectrum technology, CDMA systems employ an $(N = 1)$ frequency-reuse format.

Each CDMA base station can use the same 1.25-MHz carriers at the same time. The only change between each block of 1.25-MHz spectrum at each base station is the pseudorandom Walsh noise codes that are in use. There are a maximum of 64 allowable pseudorandom Walsh noise codes per 1.25-MHz carrier in the CDMA modulation scheme.

Prior to the widespread deployment of CDMA systems, which were mainly spurred by the broadband PCS carriers, there were concerns that CDMA systems could not handle heavy traffic loads. This caused concerns in the industry that CDMA systems couldn't handle a huge acquisition of customers in a short period of time. Nevertheless, CDMA has many distinct attributes which make it attractive to cellular and PCS providers.

Key: Theoretically, there can be nine 1.25-MHz CDMA carriers per cell. Some PCS carriers are successfully using eleven 1.25-MHz carriers per cell. Today, cellular carriers are using two to four carriers per cell. Theoretically, each of these 1.25-MHz carriers (channels) can handle 22 to 40 voice calls. However, today both PCS and cellular carriers obtain an average 12 to 16 calls per carrier.

22.5.4   Power Control

CDMA base stations control the power of all mobiles for interference reduction purposes. All mobile signals must arrive at the base station at the same power level so that the signals can be properly coded. Power control is a required operational parameter of CDMA digital systems. For example, if a mobile station that is right next to the base station is transmitting at very high power, and a mobile station 10 mi away from the base station is transmitting at very low power, the power of the mobile next to the base station is throttled down to a given level while the power of the mobile 10 mi away from the base station is raised to a given level. Power control is *necessary* to maintain system capacity. A beneficial by-product of power control is reduced power costs at the base station, as well as increased battery life in the mobile phone.

Power control exists in AMPS, TDMA, and GSM systems, but it is simply a benefit that can be utilized to make the systems perform better. Power control in CDMA systems is a *critical* item; it is absolutely required in order for the system to operate effectively.

22.5.5   Rake Receivers

CDMA systems *thrive* on multipath radio signals. Multipath fading can actually be of benefit in CDMA systems. There are four "rake" receivers within each base station transceiver and three "rake" receivers within each mobile phone, so called because they resemble a lawn rake. The function of the rake receiver at both ends (mobile and base station transceivers) is to aggregate the diversity received signals within each rake receiver. The direct signal at the rake receiver (i.e., the *strongest* signal) is combined with the multipath, reflected signals from the other two or three rake receivers to form the composite signal that is used to process the mobile call. The multipath signals are *additives* to the direct signal to obtain the cleanest, strongest signal possible.

22.5.6   Soft Handoff

"Soft" call handoffs are different from "hard" call handoffs in that a soft handoff allows *both* the original cell and one to two new cells to temporarily service a call during the handoff transition. The handoff transition is from the original cell carrying the call to one or more new cells and then to the final new cell. With soft handoff, the wireless call is actually carried by two or more cells *simultaneously*. In this regard, the analog system (and TDMA and GSM digital systems as well) provide a "break-before-make" switching function in relation to call handoff. In contrast, the CDMA-based soft handoff system provides a "make-before-break" switching function with relation to call handoff.

CDMA systems require a Global Positioning System (GPS) antenna at every cell base station. The GPS antennas synchronize all the cell sites to one timing source—the GPS. This is an absolute necessity for soft handoffs because timing is critical among the multiple sites that may simultaneously handle a call during the soft handoff process.

Analogy: The soft handoff process is analogous to a trapeze artist flying in the air between two trapeze platforms. In the analog AMPS world, the artist in the middle is transferred from the trapeze at one end to the trapeze at the other end by being flung through the air. This equates to a break-before-make switching function. Because of an RF anomaly or an error in signaling, calls could get dropped during the handoff process in analog AMPS and TDMA systems. In a CDMA world, instead of flinging the middle trapeze artist through the air, the first trapeze artist doesn't let go of the middle artist until he is sure that the opposite end trapeze artist has a firm grip on the middle artist.

Not only does soft handoff greatly minimize the probability of a dropped call, but it also makes the handoff virtually undetectable to the user. Soft handoffs are directed by the mobile telephone. As such, soft handoff is also known as *mobile-directed handoff.*

22.5.6.1 Soft Handoff Operations The sequence of events in a soft handoff is as follows:

- After a mobile call is initiated, the mobile station continues to scan the neighboring cells to determine if the signal from another cell becomes stronger than that of the original cell.

- When this happens, the mobile station knows that the call has entered a new cell's coverage area and that a handoff can be initiated.

- The mobile station transmits a control message to the MSC which states that the mobile is receiving a stronger signal from the new cell site, and the mobile identifies that new cell site.

- The MSC initiates the handoff by establishing a link to the mobile station through the new cell while maintaining the old link.

- While the mobile station is located in the transition region between the two cell sites, the call is supported by communication through both cells. This eliminates the ping-pong effect of repeated requests to hand the call back and forth between two cell sites.

- The original cell site will discontinue handling the call only when the mobile station is firmly established in the new cell.

In CDMA systems that coexist with analog/AMPS technologies (i.e., cellular carriers), hard handoffs do exist. A hard handoff in a CDMA system describes a call handoff from a CDMA carrier to an analog/AMPS RF channel. It should be noted that in CDMA systems like this, a wireless call can "step down" from a CDMA carrier to an analog channel. Yet the reverse can never happen: a wireless call can never be handed off from an analog RF channel to a CDMA-based channel.

22.5.7 Wideband CDMA

There is currently a newer version of CDMA under development, known as *wideband CDMA* (W-CDMA), sometimes known as "CDMA 2000." Instead of utilizing a 1.25-MHz carrier, W-CDMA will utilize a 5-MHz (or greater) carrier. This new technology is supposed to significantly step up the time frame by which CDMA systems will be able to offer voice, data, and at least half-rate motion video from CDMA handsets. This technology falls under the heading of third-generation (3G) wireless technology. The first generation is cellular, the second generation is all-digital (PCS) systems, and the third generation of wireless technologies are those technologies that the industry is trying to have standardized through national and international standards bodies to offer multimedia capabilities.

22.5.8 Benefits of CDMA Technology

CDMA radio technology offers the following benefits to wireless systems that implement this technology:

■ Increased *capacity* over other technologies (i.e., FDMA, TDMA) by allowing for reuse of the same (carrier) frequency in all sectors and cells. Studies show an estimated increase of about 6 to 18 times the capacity of AMPS systems. This number may rise significantly over time with the honing of current technology and the development of newer technologies such as W-CDMA.

■ *Simplified RF engineering*, due to the $N = 1$ reuse pattern in CDMA systems. This reduces the time and effort required to expand or modify CDMA systems. It should be noted that when the first CDMA systems were deployed by PCS carriers in the United States between 1996 and 1998, the carriers believed that they would need a large contingent of RF engineers to build their systems, due to the unknown nature of CDMA in a commercial environment. However, over time, the carriers realized that what they really needed were more *technicians* to drive test the systems in order to fine-tune them and tweak base station coverage.

■ Increased *performance* over the weakest link in the wireless system, the radio system. This is mainly due to the use of *rake receivers* to resolve multipath fading (Rayleigh fading).

■ *Lower transmitted power levels.* This equates to lower power bills at the base station level, and longer battery life for CDMA handsets. Power adjustments are constantly being made in the handset to reduce the amount of interference introduced to other conversations. Since a CDMA systems is noise-limited, the less interference introduced by one conversation the greater the system capacity left.

■ Greater security due to the *encoding* of CDMA signals.

■ Enhanced performance and voice quality due to the ability to accomplish *soft handoffs.*

22.6 Global System for Mobile Communications (GSM)

22.6.1 Overview

During the early 1980s, the Conference of European Posts and Telecommunications (CEPT) formed a group entitled Groupe Special Mobile (GSM) to develop a standard for the European mobile wireless market. The goals of the GSM were to provide one consistent network throughout Europe to provide roaming capabilities across the continent, security against wireless fraud, good speech quality, and ISDN compatibility. Phase 1 of the GSM specifications was completed in 1989 and the first systems were up in 1991. Commercial service began in 1992, and in 4 years there were over 35 million GSM customers being served by over 200 GSM networks.

GSM's original design specified operation at 900 MHz (in Europe). Yet the United Kingdom licensed the 1800-MHz frequencies for second-generation cellular. But instead of developing their own standards, they developed "DCS 1800" based on the GSM technology and standard. DCS 1900 is typically used in the United Kingdom, Russia, and Germany. When the FCC issued the 1900-MHz frequency range for PCS in the United States, it was again agreed that this frequency would be based on GSM and targeted for the United States and Canada. DCS 1900 was then considered the GSM standard for North America. Today, this standard is now called "North American GSM."

At the time GSM was developed, there were six incompatible cellular systems in operation throughout Europe. A mobile designed for one system could not be used with another system. This situation served as the catalyst for the development of an all-European system.

Over time, GSM has become a *worldwide* digital wireless standard, along with becoming the standard pan-European digital wireless system. This is evidenced by the fact that many American PCS carriers chose GSM as their digital radio technology standard.

Since the performance of wireless systems is restricted primarily by cochannel interference, a digital standard was sought in order to obtain improvements in spectral efficiency and increase capacity.

GSM systems use TDMA technology as their transport format (the air interface). However, the standardized approach, distinctive features [e.g., subscriber identity module (SIM) cards], and subsystem architecture of GSM make it a completely different and separate digital radio technology.

GSM operates under a strictly controlled series of standards known as the *Memorandum of Understanding* (MoU). These standards dictate how the system will operate in intricate detail. By building the system to these standards, planners could be assured that GSM systems around the world would be completely compatible.

22.6.2   Adoption of the GSM Standard

When the broadband PCS auctions concluded in the United States in March 1995, winners of the broadband PCS licenses began their decision-making processes to choose their respective digital wireless technology standards. Many of these license winners have chosen the GSM standard as their technology, such as Voicestream Wireless, Powertel, and BellSouth Mobility.

As of May 1998, GSM systems in the United States are serving almost 2 million customers in almost 1500 cities throughout 40 states and the District of Columbia, covering nearly 60% of U.S. POPs. The GSM carriers in the United States have spent billions to build out and launch their networks. They have created nearly 8000 new jobs directly, and an estimated 20,000 related jobs.

Since 1992, GSM systems have grown tremendously, to the point where there are 280 GSM networks in over 100 countries worldwide. Table 22-1 shows subscriber forecasts for GSM growth across the world.

22.6.3   The GSM Alliance

In 1997, Aerial Communications, a GSM carrier headquartered in Chicago (now purchased by Voicestream Wireless), forged an alliance among all the American GSM carriers. The purpose of the alliance was to begin development of roaming alliances between all U.S. and international GSM carriers (the network and pricing issues), development of strategic nationwide GSM marketing plans, and generally to act as a nationwide proponent of the GSM standard.

22.6.4   GSM Architecture

The GSM system offers an open architecture according to the Open Systems Interconnect (OSI) model, for layers 1, 2, and 3. This approach repre-

TABLE 22-1

Worldwide Growth
Forecast for GSM
Subscribers
(Subscribers in
Millions)

	Dec. 2000	Dec. 2001	Dec. 2002	Dec. 2003	Dec. 2004
Total GSM subscribers	331.9	435.3	537.3	631.3	715.0
GSM 900 subscribers	165.3	196.3	232.9	277.2	323.4
GSM 1800 subscribers	45.1	64.6	82.5	99.9	116
GSM 1900 subscribers	9.2	12.8	16.5	19.8	22.9
GSM 900/1800 subscribers	112.3	161.5	205.0	233.6	251.5

sents a significant departure from the previous analog systems. Also, a key benefit is afforded to GSM carriers because the architecture is a fully open system: they can go to any supplier of GSM equipment to build out their systems; they are not beholden to any one equipment supplier due to proprietary schemes.

Key: GSM systems offer an 8:1 increase in bandwidth, compared to the AMPS model. This stems from the GSM standards and the fact that full-rate vocoders are used in standard GSM systems.

22.6.4.1 GSM Subsystems The GSM network (standard) is divided into four main subsystems:

1. The base station subsystem
2. The network subsystem
3. The operations and support subsystem
4. The mobile station subsystem

The Base Station Subsystem The base station subsystem is made up of the base station controller (BSC), the base transceiver station (BTS), and the air interface. The base transceiver station consists of the antenna and the radio transceiver at the cell site. The radio transceiver defines a cell (coverage area) and controls the radio link protocols with the mobile station. The BSC is the control computer that manages many BTSs. The controller is usually housed at the MSC location, and handles which radio channels are being used by which BTSs and manages the call handoff process between BTSs. The BSC also regulates the transmit power levels of the towers and handsets. As a handset gets closer to the tower, the BSC lowers the transmitter power levels. If a handset travels away from the tower, the BSC raises the transmitter power levels. In summary, the BSC serves as a type of front-end processor for the mobile switch, the MSC. It handles much of the overhead, administrative burdens associated with frequency management, call setup, and call handoff. This allows the MSC to concentrate its processing power for what it does best: processing and switching mobile calls.

The Network Subsystem In the GSM network, the switch is the central component of the network. The mobile switching center (MSC) provides connection to the PSTN or the integrated services digital network (ISDN). The MSC also provides subscriber management functions such as mobile registration, location updating, authentication, and call routing to a roaming subscriber.

The MSC in a GSM network also houses the HLR. There is only *one* HLR in GSM networks, unless the HLR reaches capacity and a second HLR is installed. For every geographical area controlled by an MSC, there is a corresponding visitor location register (VLR). The VLR is cleared every night and starts each day with an empty database.

The security infrastructure of GSM systems is also part of the network subsystem infrastructure. This includes the authentication center, which provides the parameters needed for authentication and encryption functions. The equipment identity register (EIR) is a database used for security as well. Every GSM handset manufactured has a unique identification number known as the international mobile equipment identity (IMEI). The IMEI number of each handset is stored in the EIR. If the handset has been lost or stolen, the IMEI is placed on the blacklist of the EIR and will not work on the network. The EIR is one more way to prevent fraud on the GSM network.

The Operations and Support Subsystem The operations and support subsystem (OSS) is the command center that's used to monitor and control the GSM system. If an emergency occurs at a base station, the OSS can determine where that BTS is located, what type of failure occurred, and what equipment the site engineer will need to repair the failure.

The Mobile Station Subsystem The mobile station subsystem consists of two main components: the handset and the subscriber identity module. The handset is different from analog phones in that the identification information is programmed into the SIM card and not the handset itself. The handset's main functions are receive/transmit, and encoding and decoding the voice.

The SIM card is similar to a credit card with a microcontroller embedded into the card for the operating program and data storage. The SIM card provides authentication, information storage, subscriber account information, and data encryption. The SIM card is inserted into a handset when a user needs to make a call. SIM cards and handsets are swappable in the sense that any SIM card works with any GSM handset. Once the SIM card is plugged into the handset, the owner must enter their four-digit PIN number. The SIM card cannot be activated without this PIN number. Once the card is activated it will recall everything the subscriber has stored: speed dial numbers, messages that were received via short message service, and handset preferences.

Key: The key concept of the smart card is to associate a wireless subscription with a card, instead of a mobile phone.

In Europe, there are known instances where subscribers have had their health histories embedded into their SIM cards. When emergencies occurred, medical authorities in emergency centers simply plugged the SIM card into a GSM "reader" to obtain critical medical information on the person, such as whether they were diabetic or had high blood pressure. In the future, SIM cards will also be used as charge cards.

22.6.5   Security, Handoff, and IN

The Improved Mobile Telephone Service (IMTS) systems simply checked the MIN of the unit against a database. AMPS systems (cellular) take the security process one step further and check the MIN and the ESN of the unit against a database.

From the start, the GSM system in Europe used the *authentication* process to thwart wireless fraud. The authentication standard was included into the development of the GSM standard, and subsequently into all GSM systems that have been implemented since 1992, all around the world. North American GSM security systems use a complex series of algorithms, secret keys, and random numbers. To date, there has been no *cloning* fraud on GSM systems.

GSM also uses a distinctly digital approach to call handoff, which is known as call "handover" in Europe. In the U.S. AMPS system, call handoff is *initiated* and orchestrated by the MSC. In the GSM system, the mobile phone plays a key part in the handoff process: the *mobile phone*, not the MSC, continuously monitors other base stations in its vicinity, measuring signal strength through the MSC. The identities of the *six base stations* that are the best candidates for handoff are then transmitted back to the switch, and the network then decides when to initiate handover. GSM employs what is known in the industry as *mobile-assisted handoff.*

GSM is described as the first true instance of a wireless intelligent network (IN), as it was designed with close reference to the IN model. This is because it exhibits the following network traits:

- An open, distributed architecture
- Separation of switching and service control functions
- Full use of SS7 as the signaling infrastructure
- Clearly defined and specified interfaces
- IN structure

22.6.6   The Future of Digital Wireless Technologies

Since 1996, when the U.S. PCS carriers were selecting their digital wireless technologies from among the technologies available at that time (TDMA, CDMA, and GSM), a "religious" war has been waged between proponents of the different digital technologies. The obvious battle cry was that one or the other technology is "better" than the others. In the end, only time will tell which technology becomes the technology of the future.

In some ways, this war already has a clear loser. None of the PCS carriers who began building their networks in 1996 selected TDMA per se as their technology of choice (per se because the air interface technology of GSM is TDMA). All of these carriers selected either GSM or CDMA.

So the "war" over which technology is better has been mainly fought by proponents of these two technologies. At a high level, each technology has its benefits. GSM got a huge head start on CDMA in terms of winning acceptance from carriers the world over, mostly because it was standardized very effectively and protected the systems from fraud. It had already been adopted in well over 100 countries across the world when CDMA was first commercially launched. GSM also allows for easy roaming, even internationally. Yet it has one drawback: by definition, it can achieve only 8 to 16 times the capacity of the older, 30-kHz-channel AMPS systems. This is its main hindrance, technologically.

CDMA is touted by many industry observers as the "ultimate" in radio technology, with its state-of-the-art features (i.e., soft handoff), inherent efficiency (i.e., $N = 1$), and extremely large capacity compared to GSM systems. Many industry watchers say that CDMA is the digital technology that will take us into the future, allowing for multimedia applications on mobile handsets, data rates of 384 kb (minimum), and ever-increasing capacity gains as the technology becomes honed more and more.

The 3G standards currently under development for the third-generation wireless tend to favor the CDMA standard. But some efforts indicate there may be a merging of the best parts of both of these two technologies. Only time—and standardization—will tell.

22.7   Test Questions

Multiple Choice

1. Theoretically, how many 1.25-MHz carriers are there in a CDMA cell?
   (a) 18
   (b) 22
   (c) 9
   (d) 64
   (e) None of the above

2. The key concept of the "smart" card in GSM systems is to associate:
   (a) Multiple mobile phones for one MIN
   (b) Multiple MINs for one mobile phone
   (c) A cellular (customer's) subscription with a card, instead of a specific mobile phone
   (d) A credit card for all mobile phone users
   (e) None of the above

3. The CDMA soft handoff capability reflects:
   (a) A "break-before-make" switching function
   (b) An "ache-before-break" switching function
   (c) A "make-before-break" switching function

(d) A rake receiving switching function

(e) None of the above

4. What antifraud technology did the European GSM system employ from the start?

(a) Profiling

(b) Velocity checks

(c) Authentication

(d) RF fingerprinting

(e) All of the above

5. Which digital wireless standard began as a pan-European standard?

(a) D-AMPS

(b) CDMA

(c) GSM

(d) TDMA

(e) TDMA-6

(f) None of the above

6. Which digital radio technology employs an $N = 1$ frequency-reuse plan?

(a) GSM

(b) TDMA

(c) TDMS-6

(d) CDMA

(e) D-AMPS

True or False?

7. _____ The TDMA digital technology offers a 10:1 ratio in increased system capacity.

8. _____ Power control is not necessarily a key parameter of CDMA digital operations.

9. _____ Rake receivers choose the best signal from one of the three mobile phone or four base station receivers, and use that signal to process a CDMA digital cellular call.

23

Personal Communication Services

23.1 Overview

In the United States, Personal Communication Services (PCS) is referred to as the *second-generation* wireless service. Cellular systems are the first-generation mobile telephone service, with their unique attribute being the first deployment of frequency reuse. PCS is the second generation, being the first all-digital systems.

The FCC has defined PCS as "radio communication that encompasses mobile and fixed communications to individuals and businesses that can be integrated with a variety of competing networks." Over time, the definitions of PCS/PCN (PCN = Personal Communication Network) will be formed by the economic and technological interests of the license holders.

There are other ways to define PCS, as follows:

- *Personal mobility.* Defines the ability of a user to access telecommunications services at any terminal (wireless unit) on the basis of a personal identifier, and the ability of the network to provide those services according to the user's service profile.

- *Terminal mobility.* Defines the ability of a terminal to access telecommunications services from different locations while in motion, and the ability of the network to identify and locate the terminal.

- *Service mobility.* Defines the ability to use vertical features [e.g., custom local area signaling services (CLASS) features; centrex] provided by today's landline network to users at remote locations or while in motion. Some examples of CLASS features include call waiting, caller ID (ANI delivery), call forwarding, and automatic callback.

The main differences between PCS services and cellular services are as follows:

- PCS is an *all-digital* system from its inception. In the United States, every company that won PCS licenses had to select which type of digital wireless technology it wanted to use to build out its system: IS-136, GSM, or CDMA.

- PCS systems will allow for wireless voice, data, and video transmissions when the systems are fully developed. "Convergence" will apply to the wireless industry as well as the wireline industry.

■ PCS systems, when initially deployed, aim to have nationwide seamless roaming arrangements right from the start.

Key: Although PCS companies may have different marketing strategies than cellular companies, it is important to note that most of the fundamental concepts and attributes that are inherent to cellular system design will also apply to the PCS industry: frequency reuse, hexagon grid design, fixed network overlay, RF propagation theory, interconnection to the PSTN, roaming, etc. The core *design difference* between cellular and PCS will be that, given the same geographic coverage area, ultimately many more PCS base stations will be required to cover a market area compared to the amount of cellular base stations required (see Section 23.10, "RF Propagation and Cell Density").

Key: In contrast to telecommunication services that have evolved over the years, PCS refers to services and technologies that are *user*-specific instead of *location*-specific. Ideally, if technology advances as the FCC envisions, it may soon be possible to reach individuals at any time in any place by using a single phone number. This is referred to as "follow-me" service and is an example of a wireless intelligent network (WIN) capability such as automatic call delivery (ACD).

When fully deployed, PCS will free users from the constraints of the wireline PSTN and permit them to use the same communications device in multiple locations. The device, or terminal, may be in the form of a personal communicator, allowing voice, data, and video transmissions to occur through a small, lightweight phone.

To give this new personal form of communications technology room to grow and evolve, the FCC wisely placed few restrictions on the type of services that can be offered under the PCS banner. The FCC allocated 4 times as much spectrum for PCS as had been originally allocated for cellular communications in the early 1980s. In doing so, the intent was to provide sufficient capacity for the estimated 60 to 90 million subscribers projected to use this service by the year 2005. The PCS industry is well on its way to achieving this growth forecast. As of September 2000, there are 103,000,000 wireless subscribers in the United States. With the aggressive and innovative marketing that has been a hallmark of the PCS carriers since they came on the scene in 1996, it is likely that 35 to 50 million of these subscribers have obtained their service from PCS carriers, not cellular carriers.

23.2   Types of PCS

The PCS family of licenses is divided into two main groups: narrowband licenses and broadband licenses.

23.2.1   Narrowband PCS

By design, the 3 MHz of radio spectrum set aside for narrowband PCS is to be used primarily for data transmissions. New services expected to be provided on the narrowband PSC airwaves include advanced voice paging, two-way acknowledgment paging, data messaging, and both one-way and two-way messaging and facsimile. In July 1994, the FCC auctioned licenses for the rights to provide narrowband PCS service on a nationwide basis. Six companies paid a total of $617 million for these rights. In November 1994, the FCC auctioned off 30 regional narrowband PCS service area licenses as well.

23.2.2   Broadband PCS

In December 1994, the FCC began the auctions for broadband PCS licenses. In March 1995, the FCC concluded these auctions for 99 broadband "metropolitan trading area" (MTA) PCS licenses. Broadband PCS is intended be used for both voice and data transmissions.

Because voice transmission requires greater channel capacity than data transmission, the FCC set aside a total of 140 MHz of radio spectrum for broadband PCS usage. One of the reasons that more spectrum was allocated for PCS than cellular (140 MHz versus 50 MHz) is because of the high hopes that the FCC and the industry have for PCS services.

Key: A majority of the 140-MHz spectrum—120 MHz—has been allocated for use by licensed providers. The remaining 20 MHz of spectrum is reserved for unlicensed broadband PCS applications such as advanced cordless telephones, wireless PBXs, and low-speed computer data links.

23.3 PCS Markets

To foster the goal of encouraging the participation of both large telecommunications companies and smaller businesses in the development of PCS, the FCC issued licenses for both large metropolitan trading areas (MTAs) and smaller basic trading areas (BTAs). These divisions were drawn on a county-line basis per Rand McNally Corporation maps, and are based on the natural flow of commerce (communities of interest), in much the same way that cellular MSAs and RSAs were created.

Key: Nationwide, there are 51 MTAs and 492 BTAs.

Like cellular markets, the MTA markets are labeled according to the largest city within their boundaries. However, in contrast to cellular MSAs, MTA boundaries encompass huge swaths of territory that extend far beyond the boundaries of the largest city within the MTA border. For example, the Chicago MTA encompasses almost the entire state of Illinois, a portion of northwest Indiana, small portions of Ohio, and portions of southern Wisconsin. The Minneapolis MTA encompasses the entire state of Minnesota, western Wisconsin, eastern Montana, and part of North Dakota. This contrasts sharply with the division of cellular markets, as Illinois has many RSAs and MSAs within its state boundary.

23.3.1 Metropolitan Trading Areas

The MTAs are the largest type of PCS market, covering the largest expanses of geographic territory. Similar to the initial regulatory framework for the cellular industry where an A and B carrier were the designated market operators, there are two broadband PCS carriers operating in each MTA: an A band carrier and a B band carrier.

The MTA auctions began in November 1994 and concluded in March 1995. There are 51 metropolitan trading areas in the United States. A total of 99 MTA licenses were issued to serve the 51 MTAs across the country. (*Note:* Four "pioneer preference" licenses were already awarded for experimental systems in Washington, D.C.) The MTA licenses were assigned a total of 60 MHz of the PCS spectrum allotted by the FCC. The A and B bands each occupy 30 MHz of that spectrum, as denoted below:

A band (30 MHz): 1850–1865 MHz/1930–1945 MHz

B band (30 MHz): 1870–1885 MHz/1950–1965 MHz

23.3.2   Basic Trading Areas

There are 492 basic trading areas (BTAs) in the United States. In contrast to the initial regulatory framework for the cellular industry, there are two or more BTA carriers operating *within* each MTA, instead of adjacent to each MTA.

The BTA auctions began in 1996 and concluded in spring 1997. Although smaller than the MTAs, BTAs are still larger than most cellular MSAs and much more *expansive* than cellular RSAs. There are four distinct blocks of PCS spectrum allocated to BTAs, totaling 60 MHz of the FCC-assigned broadband PCS spectrum. They are the C, D, E, and F blocks.

Key: The total amount of licensed, broadband PCS spectrum is evenly divided between MTAs and BTAs: 60 MHz to the MTA license blocks, 60 MHz to the BTA license blocks.

There were a total of 986 licenses issued under the D and E BTA spectrum blocks. Their spectrum is allocated as shown below:

D block (10 MHz): 1865–1870 MHz/1945–1950 MHz

E block (10 MHz): 1885–1890 MHz/1965–1970 MHz

There were 493 licenses issued for both the C and F BTA spectrum blocks. The spectrum reserved for these blocks is shown below:

C block (30 MHz): 1895–1910 MHz/1975–1990 MHz

F block (10 MHz): 1890–1895 MHz/1970–1975 MHz

See Figure 23-1 for an illustration of the PCS spectrum breakdown.

Key: The broadband PCS auctions for all license blocks (A–F) (except C blocks) were concluded in spring 1997. By that time, *all* PCS auctions were complete, for both narrowband and broadband services.

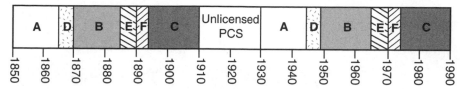

Figure 23-1 Broad-band PCS frequency spectrum plan.

- Blocks A, B are major trading areas (MTAs)
- Blocks C, D, E, F are basic trading areas (BTAs)
- Unlicensed PCS is nationwide

23.4   Licensing Mechanisms

In the past, the FCC issued licenses for radio spectrum without requiring direct payment to the government. In many instances, cellular companies were able to make windfall profits by selling their licenses soon after they were awarded. Seeing the opportunity to put an end to such profiteering and raise revenues at the same time, Congress in 1993 mandated that the FCC release licenses for PCS through an auction procedure and issued rules governing postauction sale of the licenses as well.

23.5   The PCS Auctions

In March 1995, the U.S. government obtained over $7 billion by auctioning off what amounts to the right to project radio waves at a certain frequency through the air. This amount of money was paid just for PCS licenses at the conclusion of the auctions in March 1995! This didn't include the huge, multibillion dollar cost required to build out the PCS systems themselves.

Never before in our nation's history has radio spectrum been sold; it has always been granted by the FCC. In some circles, the auctioning of national radio spectrum is viewed as a dangerous precedent. This is because it amounts to the selling of a national resource. Some say that it's tantamount to selling off a national park, wildlife refuge, or national museum. Only time will tell whether the government's shrewd but questionable audacity in this venture will apply to other national business topics in the future. See Figure 23-2 for a listing of broadband PCS licensees, market licenses they won, and prices paid "per POP" in those respective MTA markets.

Broadband PCS auction winners and the markets they won

COMPANY	CONTACT	MARKET	BID	PER POP
American Portable Telecommunications Inc. 30 N. LaSalle St. No. 4000 Chicago, Ill. 60602 (202) 467-5700	George Y. Wheeler	Minneapolis	$36.6 million	$6.11
		Tampa	$89.8 million	$16.57
		Houston	$83.9 million	$16.16
		Pittsburgh	$31.7 million	$7.72
		Kansas City	$23.6 million	$8.10
		Columbus	$22.2 million	$10.34
		Alaska	$1 million	$1.82
		Guam	$141,837	$.81
Ameritech Wireless Communications Inc. 30 S. Wacker Drive Chicago, Ill. 60606 (708) 248-6128	Steven Engelkrohl	Cleveland	$87 million	$17.59
		Indianapolis	$71.1 million	$23.56
AT&T Wireless PCS Inc. 1150 Connecticut Ave. N.W. 4th Floor Washington, D.C. 20036 (202) 223-9222	Cathleen A. Massey	Chicago	$372.8 million	$30.88
		Detroit	$81.2 million	$8.12
		Charlotte	$66.6 million	$6.63
		Boston	$121.7 million	$12.87
		Philadelphia	$81 million	$9.07
		Washington D.C.	$211.8 million	$27.23
		Atlanta	$198.4 million	$28.58
		Cleveland	$85.9 million	$17.36
		Cincinnati	$41.9 million	$8.89
		St. Louis	$118.8 million	$25.48
		Richmond	$33.7 million	$8.75
		Puerto Rico	$56.9 million	$15.70
		Louisville	$49.3 million	$13.85
		Phoenix	$78.3 million	$22.32
		Buffalo-Rochester	$19.9 million	$7.15
		Columbus	$22.3 million	$10.39
		El Paso-Albuquerque	$8.6 million	$4.08
		Nashville	$15.8 million	$8.95
		Knoxville	$10.6 million	$6.18
		Omaha	$4.6 million	$2.80
		Wichita	$4.4 million	$3.91

COMPANY	CONTACT	MARKET	BID	PER POP
PCS PrimeCo L.P. 1818 N St. N.W. No. 800 Washington, D.C. 20036 (202) 293-4960	Kathleen Abernathy	Chicago	$385.1 million	$31.90
		Dallas-Fort Worth	$87.5 million	$9.03
		Tampa	$99.3 million	$18.33
		Houston	$82.7 million	$15.93
		Miami	$126 million	$24.53
		New Orleans	$89.5 million	$18.17
		Milwaukee	$86 million	$18.94
		Richmond	$33 million	$8.59
		San Antonio	$32 million	$17.39
		Jacksonville	$44.5 million	$19.56
		Honolulu	$21.7 million	$19.56
PhillieCo L.P. 1850 M St. N.W. No.1110 Washington, D.C. 20036 (202) 828-7452	Jay C. Keithley	Philadelphia	$85 million	$9.52
Poka Lambro Telephone Cooperative Inc. 11.5 miles north of Tahoka, Texas on U.S. 87 Tahoka, Texas 79373 (202) 296-8890	Stephen G. Kraskin	Spokane-Billings	$5.9 million	$3.05
		Guam	$107,000	$.61
Powertel PCS Partners L.P. 421 Gilmer Ave. Lowell, Ala. 36863 (205) 644-9440	James A. Murrel	Memphis	$43.2 million	$12.46
		Birmingham	$35.3 million	$10.87
		Jacksonville	$46 million	$20.22
South Seas Satellite Communications Corp. c/o 25 N. Stonington Road South Laguna, Calif. 92677 (714) 499-4469	Lynnea Brylund Dalton Marcus K. Dalton	American Samoa	$214,555	$4.57

Company	Contact	Market	Value	Price per pop
BellSouth Personal Communications Inc. 3353 Peachtree Road No. 400 North Tower Atlanta, Ga. 30326 (404) 841-2040	Rebecca A. Jackson	Charlotte Knoxville	$70.9 million $11.1 million	$7.27 $6.47
Centennial Cellular Corp. 50 Locust Ave. New Canaan, Conn. 06840 (908) 223-6464	Rudy J. Graf	Puerto Rico	$54.7 million	$15.09
Communications International Corp. 717 W. Sprague No. 1600 Spokane, Wash. 99204-0466 (509) 623-2028	Neil S. McKay	American Samoa	$228,001	$4.85
Cox Cable Communications Inc. 1320 19th St. N.W. No. 200 Washington, D.C. 20036 (202) 857-2824	Laura H. Phillips	Omaha	$5.1 million	$3.06
GO Communication Corp. 2550 Denali St. No. 1000 Anchorage, Alaska 99503-2781 (907) 266-5647	Richard Dowling	Alaska	$1.7 million	$3.00
GTE Macro Communications Corp. 245 Perimeter Center Pkwy 3 Bsg Atlanta, Ga. 30346 (202) 463-5292	Carol Learem Bjelland	Atlanta Cincinnati Denver Seattle	$184.7 million $42.7 million $64.5 million $106.4 million	$26.60 $9.06 $16.62 $27.79
Pacific Telesis Mobile Services 4420 Rosewood Drive Building 2, 4th Floor Pleasanton, Calif. 94588 (510) 227-3015	Mike Patrick	Los Angeles San Francisco	$493.5 million $202.2 million	$25.78 $17.00
Southwestern Bell Mobile Systems Inc. 17330 Preston Road No. 100-A Dallas, Texas 75252 (214) 733-2000	Paul L. Boris	Memphis Little Rock Tulsa	$43.2 million $12.7 million $17.6 million	$12.46 $6.21 $16.02
Western PCS Corp. 3301 120th Ave. N.E. No. 200 Bellevue, Wash. 98005 (206) 635-0300	John W. Stanton	Portland Des Moines Salt Lake City El Paso–Albuquerque Oklahoma City Honolulu	$34.2 million $22.1 million $45.8 million $8.6 million $11.1 million $22.4 million	$11.16 $7.35 $17.82 $4.08 $5.92 $20.18
WirelessCo L.P. 1850 M St. N.W. No. 1100 Washington, D.C. 20036 (202) 828-7453	Joy C. Keithley	New York San Francisco Detroit Dallas–Fort Worth Boston Minneapolis Miami New Orleans St. Louis Milwaukee Pittsburgh Denver Seattle Louisville Phoenix Birmingham Portland Indianapolis Des Moines San Antonio Kansas City Buffalo–Rochester Salt Lake City Little Rock Oklahoma City Spokane–Billings Nashville Wichita Tulsa	$442.7 million $206.3 million $86.1 million $80.4 million $127.1 million $39.7 million $131.7 million $93.9 million $114.43 million $85 million $20.7 million $44.4 million $105.2 million $46.6 million $75.6 million $35.6 million $34.1 million $70.4 million $21 million $54.4 million $23.6 million $18.9 million $46.2 million $12.3 million $13.1 million $6.2 million $16.4 million $4.9 million $16.6 million	$16.76 $17.37 $8.61 $9.12 $13.44 $6.63 $25.64 $19.07 $24.51 $18.73 $7.00 $16.60 $27.48 $13.10 $21.54 $10.97 $11.16 $23.34 $7.00 $18.21 $8.11 $6.00 $17.95 $6.01 $7.00 $3.32 $9.26 $4.36 $15.32

Figure 23-2 Broadband PCS auction winners and the markets they won. (Source: RCR Magazine, March 1995.)

23.6 PCS Partnerships

Many industry sources speculate the cost to fully build out the PCS systems will be *double* what the licenses themselves cost. Because of the huge cost of obtaining the PCS licenses and the subsequent massive investment required to build out these systems, many large telecommunications companies formed partnerships before going into the auctions. Many of the PCS bidders underwent name changes from 1995 to 1999. Some of the more prominent PCS partnerships that were formed are as follows:

■ *Sprint Spectrum.* A company known as WirelessCo L.P. at the time the auctions began in late 1994 changed its name to Sprint Telecom Venture (STV) in 1995. This company is now finally known as Sprint PCS. It is a partnership of the nation's third-largest long distance carrier and three major cable television companies: Cox Communications, Telecommunications Incorporated (TCI), and Comcast Communications. Industry sources speculate that there were shrewd reasons behind Sprint's choice to partner with cable TV companies: Sprint may be able to make extensive use of the cable companies' existing network infrastructures for the PCS fixed network overlay. Sprint may also make use of the many telephone poles used by its cable TV partner companies to mount equipment and antennas for PCS base stations.

■ *Primeco Personal Communications.* This partnership, initially known as PCS Primeco, consists of several of the Regional Bell Operating Companies cellular companies, and a cellular carrier spun off from Pacific Telesis. They are Nynex Mobile, U.S. West's New Vector cellular company, Bell Atlantic Mobile Systems (BAMS), and Airtouch Cellular.

Other Bell companies and large independent telephone companies also bid for and won MTA (broadband) PCS licenses:

■ BellSouth produced a PCS subsidiary company, BellSouth PCS.

■ Telephone and Data Systems, parent company of United States Cellular, created a subsidiary company known as American Portable Telecommunications (APT). The company changed its name in November 1996 to Aerial Communications, as part of a move to gain brand recognition.

■ Ameritech bid for and won the licenses for only two MTAs, which happen to be in its own operating territory: Cleveland and Indianapolis.

- GTE, the nation's largest independent telephone company, bid for and won four broadband PCS licenses under the name GTE Macro: Atlanta, Cincinnati, Denver, and Seattle.
- Pacific Telesis (PacTel) Mobile Services won the MTA licenses for Los Angeles and San Francisco.

The following GSM companies were little known at the time of the auctions, but are now fairly well known in the industry. The regions where they obtained licenses and built out their systems are indicated:

- PowerTel—Southeastern United States
- Omnipoint—Eastern seaboard and portions of the midwest (e.g., Indianapolis)
- Western Wireless—western and northwestern United States

The names, ownership, and makeup of many wireless companies changed in the late 1990s after the PCS licenses were issued. Most of these changes were due to mergers and buyouts, as the American telecommunications industry was not immune to the merger mania that swept the world in the 1990s. Some examples are listed below:

- Aerial Communications, a GSM operator and license holder of eight MTA markets, was purchased by Voicestream Wireless, based in Bellevue, WA.
- Voicestream Wireless also purchased Omnipoint Communications, a GSM operator in the eastern and northeastern United States. These two purchases reflect Voicestream's efforts to attain a nationwide GSM footprint to compete with AT&T Wireless and Sprint PCS, since those carriers have nationwide coverage and simple rate plans in TDMA and CDMA systems, respectively.
- The partnership of companies that formed PrimeCo PCS has stated its intent to break the company up. It is still uncertain how it will sell or spin off its 11 MTA markets nationwide. Industry speculation states that one of the original PrimeCo partners may purchase the other partners' shares; or that each market will be sold individually.
- Ameritech Wireless is now SBC Wireless, since SBC purchased Ameritech in 1999. It should be noted that as a condition of the merger, some of the Ameritech wireless markets had to be sold or spun off due to the fact that Ameritech and SBC owned both the A and the B band licenses in several cellular markets. Most of these markets were in Illinois—including the Chicago MSA—and GTE

purchased those properties. GTE's merger with Bell Atlantic went through at around the same time, and *that* merged entity's name is now Verizon Communications. This means that the former Bell Atlantic Mobility Systems (BAMS) also does not technically exist anymore. It is also known as "Verizon Communications."

■ Pacific Telesis Mobile Services is now SBC Wireless, since SBC purchased Pacific Bell in 1997.

■ SBC Wireless and BellSouth Wireless entered into a partnership in August 2000 in order to allow for easy roaming across the southern United States.

23.7   Funding

Most companies that won PCS licenses relied on their parent companies to fund them through the start-up phase, until they could activate their systems, begin selling service, and obtain positive cash flow. It is projected to take at least 4 to 10 years to obtain positive cash flow, because of the huge cost of obtaining the licenses and building the systems. This depends on the carrier and the market. It is easier for the partnership companies to rely on their parent companies to absorb the cost for the PCS start-up phase, because of the parent companies' large size and huge revenue base. There is only one communications company that could bear these costs alone, because of its massive size and revenue sources: AT&T Wireless.

The *total* cost to deploy a specific PCS system (MTA or BTA) will depend on the market in question. The cost to buy or lease land will vary among the markets, as will overall construction costs. PCS companies, even more than cellular carriers, will have to overcome the tower zoning moratorium that exists in many localities. Some companies that bid on licenses went bankrupt before they could even get a start. The most prominent example to date is the C block winner in the Chicago market, Pocket Communications.

23.8   Measurement of Market Values and the Big Winners

The Chicago MTA turned out to be the most expensive market to bid for, per POP. Primeco Personal Communications paid $31.90 per POP, for a total of $385 million, to obtain one of the MTA licenses. AT&T Wireless was the other winner for the Chicago market, paying $372 million.

Sprint Spectrum and AT&T Wireless won the largest number of licenses out of all the companies bidding nationwide, each spending over $2 billion for their MTA licenses.

Other notable license winners were Primeco Personal Communications with 11 licenses, Aerial Communications with 8 licenses, and Western Wireless with 6 licenses.

23.9   FCC Restrictions

PCS licensees are *not* restricted in the number of MTAs or BTAs they can operate. However, they are restricted in the *amount of radio spectrum* they can own and operate in any one service area.

Key: No wireless company can control more than 50 MHz of PCS-allocated spectrum in any one market.

23.9.1   Restrictions Placed on In-Market Cellular Companies

A cellular company bidding for a broadband PCS service area (MTA) that overlapped its existing cellular service area was restricted in the amount of spectrum for which it could bid. The FCC ruled that if 10% or more of the population of the PCS service area is within the cellular company's existing service area, the cellular company could bid only for one of the smaller BTA broadband spectrum blocks (10 MHz). One option for the cellular company was to sell its cellular markets in cases where its existing markets overlapped 10% or more of a given MTA service area's POPs. This option is proven with the spinoff of Sprint Cellular from Sprint Corp.

Sprint Cellular became its own company in the first quarter of 1996, completely separate from Sprint Corp. It also was required to change the name of the company, which became 360 Degrees Communications. Alltel Mobile has since purchased 360 Degrees Communications, in 1998. This divestiture was necessary because Sprint's PCS consortium won the second-largest amount of nationwide broadband PCS licenses next to AT&T Wireless. The large amount of geography that Sprint PCS's MTAs cover overlapped most of Sprint's existing cellular markets, forcing Sprint to spin off Sprint Cellular.

23.9.2   Postauction Requirements and Construction Permit Issues

To discourage successful PCS bidders from making a quick profit, the FCC requires that PCS licensees adhere to strict construction requirements. The FCC also prohibits certain licensees from selling their operating rights for a given period of time, or the FCC will impose stiff penalties. After winning a PCS license and making the required down payment, licensees are expected to provide service to a certain percentage of the population of the service area within 5 years. This requirement is similar to the cellular service 5-year CP requirement. Licensees of the 30-MHz broadband PCS spectrum blocks (MTAs) will be required to provide coverage to at least one-third of the population of their markets within 5 years, and to two-thirds of the population within 10 years.

Licensees of the BTA 10-MHz broadband blocks have less stringent requirements. The FCC mandates coverage to at least one-fourth of the population of their service area within 5 years, or to at least demonstrate to the FCC that they are providing "substantial" service to the populace in their service areas.

23.10   RF Propagation and Cell Density

Because radio propagation at *higher* frequencies fades more quickly due to free-space loss, many more cells will need to be constructed overall to build out PCS networks when compared to the construction of cellular networks. Most wireless engineers state that the required ratio of PCS cells to cellular cells will need to be 3:1 (average) to deliver comparable coverage in PCS markets. In light of this assessment, many industry sources say that PCS cell sites will be nearly ubiquitous in the long run: appearing on street lights, traffic lights, church steeples, motel signs, on utility poles, maybe even ultimately on houses and garages.

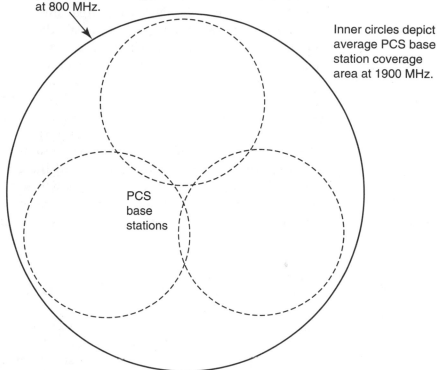

Figure 23-3
Projected increase in PCS base station density compared to cellular coverage: 3:1 ratio (average).

Large circle depicts average cellular (AMPS) base station coverage area at 800 MHz.

Inner circles depict average PCS base station coverage area at 1900 MHz.

PCS base stations

Key: It is estimated that 100,000 antenna sites (PCS base stations) will need to be installed nationwide with the rollout of PCS services. Many of these sites will place antennas on existing towers or on tall buildings, but nearly 30% of these sites will be new tower structures more than 200 ft tall. See Figure 23-3 for a comparison of PCS base station requirements to deliver coverage equivalent to cellular base stations.

23.11   PCS Site Zoning Issues

Finding suitable tower and antenna sites and obtaining zoning approval for them ranks as a main challenge for PCS providers.

A paradox is at play concerning the American public's feelings on wireless service. Although many citizens acknowledge the positive attributes of wireless service, they do not want antennas and towers in their "back-yards." The resulting zoning battles can be both expensive and time-consuming for wireless carriers. On this front, the CTIA has worked to help PCS providers. The NIMBY (not in my backyard) philosophy has taken root in many communities to thwart the construction of towers to support the deployment of wireless base stations. The zoning boards of many municipalities now require wireless carriers to (attempt to) colocate onto existing towers as a prerequisite for zoning approval of new PCS base stations. The CTIA has also asked the FCC to establish standardized guidelines for local zoning boards and governments to follow.

In 1996, the president of the CTIA, Thomas Wheeler, asked President Clinton to assist by having government grounds and areas available for antenna location. The President supported the CTIA in this matter by having the United States Postal Service designate a Postal Service point-of-contact for the carriers, who is responsible for helping PCS carriers find post office sites to locate PCS base station antennas.

23.12   Interconnection Issues

Since interconnection to the PSTN can have one of the biggest impacts on a wireless carrier's profit and loss, many PCS providers have been concerned about obtaining *true* cocarrier status with regard to interconnection. Forecasts are projecting that more calls are expected to *terminate* on the wireless networks in the PCS world, as compared to the 30 to 35% of calls that terminate on cellular networks today. This is why the FCC included a rule on the interconnection policy of reciprocal compensation in the Telecom Act of 1996, where each carrier (PCS and landline telcos) would compensate the other carrier for terminating traffic.

23.13   PCS Marketing and Service Offerings

The marketing departments of the PCS carriers are trying to bundle their service offerings in a manner which makes the service more attractive to potential customers than cellular services. There is also no extra charge for some of the services (e.g., voice mail and caller ID). Examples of such services are as follows:

- Short message service (SMS) for transmission of short e-mail messages directly to the PCS handset
- Voice mail with message waiting notification
- Caller ID
- Paging services
- Internet access
- Automatic callback
- Call forwarding

Many PCS carriers are also employing a strategy known as *big minutes* to compete with cellular carriers and garner new customers in a short time frame. Big minutes describes the marketing rate plans that offer large amounts of network minutes-of-use for a relatively low flat monthly fee. For example, PCS carriers will advertise "$50 for 1000 *anytime* minutes per month." The "anytime" part of this offer is significant because the *cellular* carriers have "free minutes" offers, but historically have restricted the customers to the free minutes during off-peak hours, which are usually after 7:00 P.M. and on weekends. This tactic is employed in these marketing campaigns so the PCS carriers can differentiate themselves from cellular carriers. Most people choose to use their phones more often during hours *prior* to 7:00 P.M. Big minutes rate plans are also an attempt by the PCS companies to convince many customers that the only phone they need is their wireless phone. The goal is to displace the need (or perceived need) for a *wireline* phone.

23.14   Competition with Cellular

The licensed providers of PCS services are expected to compete with cellular carriers by offering lighter-weight phones, a denser concentration of cells (better coverage), and lower airtime rates (cost per minute) than currently charged by cellular carriers.

Key: The concept of lower airtime rates *may* seem a little hard to believe if PCS providers will need to build approximately three PCS cells for every one cell the cellular carriers have built. However, the concept is not far-fetched because PCS base stations are more technically advanced, more compact, and therefore less expensive than cellular base stations. PCS base stations can even come in the form of stand-alone, weatherproof cabinets that are 7 ft long, 5 ft tall, and 2.5 ft in depth.

With MTA licensees and BTA licensees, there will be many wireless carriers competing with the currently entrenched cellular carriers. That means that the general public may eventually have up to six wireless carriers to select from as their wireless service provider. The competition with cellular providers is intense to say the least, as the PCS companies quickly seek revenue streams to compensate for the huge cost of the PCS licenses and the subsequent system buildouts. In anticipation of the competition with PCS carriers, in the 1994–1995 time frame, many cellular carriers began taking the word *cellular* out of their names, and replacing it with the word *wireless* in order to promote the concept that there is no difference between cellular and PCS. This also ended up confusing the consumer in the end.

23.15   Digital Technology Selection

The winners of the broadband MTA licenses spent the last half of 1995 determining which digital technology they would select so they could begin building out their networks. Their selections varied among the multiple choices available (TDMA, GSM, CDMA), including some technologies that were popular yet still not 100% developed (tested) for commercial applications at the time (i.e., CDMA).

This was a very serious decision for all the carriers who won licenses, because there were far-reaching implications:

■ The carriers had to select a technology that they thought would be "future-proof"—one that would thrive and remain viable for a long time (i.e., CDMA).

■ They had to select a technology that was standardized, and would be technically "popular" to enable them to implement widespread roaming capabilities, even internationally (i.e., GSM).

■ They had to select a technology that was proven and already had a large embedded base (i.e., TDMA).

It is easy to see that the license winners had difficult technology decisions to make. But it is interesting to note that few, if any, of the PCS license winners selected IS-136 (TDMA) as their digital technology of choice. This is because IS-136, when compared to GSM and CDMA, has little to offer in terms of being a digital wireless technology of the future.

23.16   Dual-Mode PCS Handsets

PCS providers will be required to provide dual-band, dual-mode handsets to their customers in order to compete with and coexist with their cellular counterparts. This means the handsets would operate in both the 800- and 1900-MHz frequency bands, and in analog or digital mode. Since each PCS carrier may be building its markets out at different paces, or certain portions of its markets at different paces, PCS phones must be able to convert to an analog (cellular) mode for AMPS roaming purposes. PCS providers will also need to develop roaming agreements with cellular carriers so that PCS customers can roam onto cellular systems. There will ultimately be a need for a PCS handset that can operate in several digital modes as well as the analog mode.

23.17   Roaming and Digital Interoperability

The main problem with so many digital radio system technologies being fielded by the PCS carriers is that, if a PCS customer were to roam to a nearby market, that customer may need to have a triple-mode or quadruple-mode handset in order to obtain operating access to a "visiting" system. By the year 2002 any given locality in the United States could have up to four different air interface technologies in operation: analog/AMPS, TDMA, CDMA, and GSM.

During the first 10 to 12 years of the cellular industry's existence in the United States, it was relatively easy to place calls if you roamed into another cellular market. You could be fairly sure that making a call would not be a major hassle, especially over the last few years with the deployment of national, SS7-based networks and roaming agreements which allow for seamless automatic call delivery to roaming subscribers.

During the first few years of PCS deployment, this may not necessarily be the case. PCS will offer a different scenario concerning roaming because of the deployment of three distinct digital technologies, and the situation of possibly needing triple- and maybe quadruple-mode phones must be addressed to avoid customer aggravation when roaming. An industry leader, Motorola, has addressed this issue. In 1999, Motorola announced it had developed a chip set that could operate in all three of the digital wireless modes: IS-136 (TDMA), CDMA, and GSM. These phones have not come to market yet (circa 2000), but it is only a matter of time before they do. The deployment of triple- or quadruple-mode wireless phones into the marketplace will change the face of the wireless industry forever, as roaming anywhere, on any system, regardless of the technology used by a wireless subscriber's "home" system carrier, will become a reality. The estimated time frame for this evolution to develop and take root on a widespread basis is the year 2004.

23.18   Microwave Relocation

23.18.1   Overview

There has been a large obstacle to the process of implementing PCS systems nationwide. The spectrum between 1850 and 1990 MHz that was allocated to PCS carriers by the FCC was occupied by more than 474 microwave "incumbents" operating more than 5000 point-to-point microwave radio systems nationwide. The microwave incumbents include public, private, and semiprivate entities such as utilities, railroads, public safety administrations, governments, and cellular carriers (fixed network). These systems sometimes spanned multiple states, and represented rural, urban, and suburban interests.

The FCC established rules for sharing this spectrum in 1991, stating that the newly licensed PCS operators could not interfere with the currently operating microwave systems.

The FCC rules established broad guidelines for eventually relocating these microwave users to new spectrum or alternative technologies—to "comparable facilities." The guidelines concerning relocation of microwave facilities are as follows:

- For incumbent microwave facilities that were licensed *before* January 1992, the incoming PCS operator will have to bear the cost of moving the incumbent to comparable facilities.

- For incumbent microwave facilities that were licensed *after* January 1992, the incumbent operator of the microwave facility will have to bear the cost of relocation itself. This is because the incumbent operator was warned ahead of time that it should not implement microwave systems in the 1.9-GHz (1900-MHz) frequency range.

The rules governing the burden of the cost of relocation were set up in this manner because, in late 1991, the FCC orders gave fair warning to all parties who were applying for microwave licenses at that time. The orders stated that the 2-GHz band spectrum was being set aside for PCS services, to be auctioned off in several years. So around that point in time, microwave license applicants were made aware that they would most likely have to relinquish their microwave license at some time in the near future. They then installed 2-GHz systems at their own risk. Although some microwave incumbents were aware of this proposed resolution for sharing spectrum, many others received their introduction to the process when a PCS licensee initiated a call to them to discuss relocation.

23.18.2   Relocation Costs

The cost to relocate microwave systems is a major issue relating to the deployment of PCS systems. This is because the cost to relocate a single microwave path could easily run well into the *hundreds of thousands* of dollars, depending on the type of system in place. Some incumbent microwave operators stooped to forms of extortion when incoming PCS operators requested relocation of the incumbent's microwave system. For example, if the estimated cost to relocate the incumbent's system was around $400,000, the incumbent may have demanded a "comparable facility" that's actually worth $850,000! The FCC (and the CTIA) received numerous complaints about this situation. In 1996 the FCC filed a *Notice of Proposed Rulemaking*, stating its desire to implement a cost-sharing plan for the microwave relocation process. Newly proposed rules would create a clearinghouse mechanism to administer the costs and obligations of PCS licensees that benefit from the spectrum-clearing efforts of other licensees. However, the CTIA stated that such rules should also permit parties to individually negotiate cost-sharing arrangements. In addition, the CTIA stated that the bargaining power of microwave incumbents and PCS licensees must be equalized by creating incentives for microwave incumbents to relocate during a voluntary negotiation period.

23.18.3   The Relocation Process

Every relocation of a microwave path must be addressed differently, because incumbents have tremendously varied needs, interests, resources, and incentives. Because this is a unique process, negotiations between PCS licensees and microwave incumbents were often complex, but always interesting. In other words, the negotiations had the potential to be hostile. Not all microwave paths occupying the 2-GHz spectrum were an immediate interference problem. PCS licensees have been approaching this issue in different ways, depending on their market strategy:

■ Some PCS licensees, eager to deploy their systems, adopted a first-to-market strategy, and were aggressive in beginning discussions with incumbents.

■ Other PCS licensees took a more deliberate approach, concentrating their resources on prioritizing the deployment of multiple markets, or various areas within a market.

As with microwave incumbents, resources, strategies, and goals vary widely among PCS licensees, contributing to the complexity of the relocation process. Because microwave relocation is a new process, every resolution that was reached set an industry precedent. Many microwave incumbents and PCS licensees are acutely aware of this, and they are attempting to resolve contentious issues through creative and mutually beneficial solutions. The parties that have been able to do this have followed some basic guidelines to ensure their success:

■ *Know the issues and get the resources.* Many PCS licensees have taken the time to educate themselves on all the potential stumbling blocks associated with microwave relocation and have taken necessary steps to ensure these problems are avoided whenever possible.

■ *Find and realize economies.* Some PCS licensees have attempted to realize economies of scale by analyzing where there may be overlapping relocation discussions and holding these simultaneously.

■ *Develop comprehensive relocation strategies.* PCS licensees should investigate not only the expectations of each microwave incumbent, but also their own parameters and priorities for negotiation. They need to determine whether incumbent partnership opportunities are of internal interest, such as providing PCS services together, sharing microwave towers, or leasing excess system capacity.

■ *Manage the process.* Many PCS licensees have anticipated the magnitude of the relocation process, and they have created and assigned special teams devoted solely to managing this process.

23.19   Deployment Timelines and Market Launches

The time needed to build the PCS networks could make the difference in gaining an edge on competitors, with most networks requiring at least a year to deploy, according to industry players.

Some PCS systems were on line in the third quarter of 1996 (e.g., Primeco PCS in Chicago). Most other PCS carriers launched service in Spring 1997. At launch, the PCS carriers developed and distributed detailed maps showing customers where coverage was at launch, as well as areas where they planned to have expanded coverage within 1 year. By late 1997, serious competition between cellular and PCS providers had begun.

By late 1996, some cellular carriers, anticipating the competitive onslaught of the PCS carriers, began offering attractive service packages to customers. However, there's a catch: They required customers to sign 2-year contracts. That way, the customer couldn't break the contract to migrate to a PCS carrier without facing stiff penalties.

23.20   Marketing Issues

It will be crucial for the PCS carriers to build high-quality networks with high-quality customer service so they can build customer loyalty from the start. This factor will be heavily dependent on finding suitable sites for PCS base stations. The second most important thing will be to have competence in your company. This is especially critical in the PCS industry because there has never been an industry (segment) that is so capital-intensive with so many competitors, *from the start!* One industry source offered this advice to PCS carriers: "Have well-thought-out engineering plans. Site acquisition and construction take longer than you think. Start interconnection *now.*" The challenge for PCS service will be to sell not only service, but *value.*

23.21   Test Questions

Multiple Choice

1. Why did the FCC issue licenses for both MTA-type markets and BTA-type markets?
   (a) This design was implemented randomly
   (b) To foster close cooperation with cellular companies
   (c) To foster the participation of both large telecommunications companies and smaller businesses
   (d) To ensure there were many PCS markets
   (e) All of the above

2. What are the three types of mobility that PCS systems will offer?
   (a) Personal mobility
   (b) Carrier mobility
   (c) Satellite mobility
   (d) Terminal mobility
   (e) Service mobility
   (f) a, b, and c above
   (g) a, d, and e above
   (h) a, c, and d above

3. In contrast to telecommunications services that have evolved over the years, PCS refers to services and technologies that are:
   (a) Location-specific instead of user-specific
   (b) MTSO-specific
   (c) LEC central office–specific
   (d) User-specific instead of location-specific
   (e) a and c only
   (f) None of the above

4. The FCC rules established broad guidelines for eventually moving microwave users to:
   (a) Fiber-optic facilities only
   (b) Infrared technologies
   (c) Comparable facilities
   (d) None of the above

5. The FCC has ruled that if _____% or more of the population of a PCS service area is within the cellular company's existing service area, the cellular company may bid for only one of the smaller broadband spectrum blocks (10 MHz).

(a) 50

(b) 20

(c) 15

(d) 10

(e) None of the above

True or False?

6. _____ Because of the nature of RF propagation in the PCS spectrum, PCS carriers will need to install far fewer cells overall than cellular carriers have had to install.

7. _____ Infrastructure costs for PCS carriers—per base station—will be much higher than they are for cellular carriers.

8. _____ For incumbent microwave facilities that were licensed after January 1992, the incoming PCS operator will have to bear the cost of moving the incumbent to comparable facilities.

24

Wireless Data Technologies

24.1 Overview

The use of wireless services is predicated on one main idea: we need to be as available and productive as possible, regardless of location. With the increased amount of digital information available today, mobile communication systems have been required to keep pace by including mechanisms for transmitting data. Today, these applications include the transmission of electronic mail, faxes, web browsing, and access to real-time stock information. Tomorrow's applications may include video conferencing, location-specific advertising (facilitated by new locator technologies that use triangulation), as well as personal entertainment services.

All of these applications can be broken down into three main categories:

1. Query/response applications
2. Batch file applications
3. Streaming data applications

Examples of query/response applications include (remote) telemetry such as remote power readings, vending machine management, copier management, monitoring of burglar alarm systems, and transfers from other sensing devices. Query/response applications also lie at the heart of most information services and wireless e-mail access.

Batch file applications are used to send large data files that are not time-sensitive. Typical applications include transmission of electronic documents (i.e., faxes), and shipping manifests.

Streaming data applications contain time-sensitive information, but do not require guaranteed delivery. This application area has the potential to explode by 2004 given the possibility of new services such as delivery of video programming, video conferencing, and streaming audio services.

The tools for accessing wireless data systems are also in a major state of change. Originally accessed by specialized cellular equipment slaved off of a laptop computer or personal digital assistant, wireless data systems can now be accessed using browsing software that's integrated right into wireless phones. New "body LAN" technologies now allow devices to automatically communicate without cables. As a result of these new technologies, changes made to the calendar on a personal digital assistant can be automatically updated on a laptop computer located in the person's nearby briefcase, and automatically sent to the corporate calendar service via a nationwide packet network.

24.2 Wireless Data Use Paradigms

Three paradigms exist for wireless data use:

1. *Full user mobility.* Allows for session handoff while moving—during automobile travel, for example.

2. *Portable wireless data.* Allows access to a data service while in a given coverage area, using a laptop or palmtop computer and a wireless modem. This functionality doesn't offer full mobility or session hand-off during on-line sessions. Access to the Internet while moving at a relatively high rate of speed will result in a dropped connection at the edge of the coverage area.

3. *Fixed wireless data.* Service is offered to an office or home through large customer-premise antennas. In this context, data throughput speeds can be expected to achieve rates up to 45 Mbps (DS3 rate).

Within these paradigms, there are two media access methods in wireless data communications:

- Circuit-switched access
- Packet-switched access

Circuit-switched wireless access establishes a standard air link ("air interface") connection to the nearest cell site, to a switch (MSC or peripheral), and then into the public switched telephone network (PSTN). This is accomplished by using modem contention and data transmission, usually employing Microcom Networking Protocol class 10 (MNP 10) or Enhanced Throughput Cellular (ETC) protocols. With the exception of the noise and distortion inherent in the air link, the circuit-switched data call is no different from a landline-based circuit-switched data call (i.e., dial-up Internet access). Of course, the protocols that are used to negotiate, transmit, and receive user data in a wireless context are designed specifically for this purpose as well. Figure 24-1 depicts a typical wireless circuit-switched data call.

Figure 24-1
Circuit-switched data call.

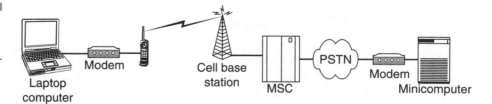

Circuit-switched wireless applications can run the spectrum of various remote access applications, including on-line services, e-mail, fax, local-area network (LAN) access, and real-time file transfers.

Packet-switched wireless access establishes a session, but does not have a dedicated bearer channel for the user data like circuit-switched data. Instead, the wireless access device is able to request time slots to send data as packets are generated by the source application. Likewise, the base station will send data addressed to the wireless access device on a time slot when the base station receives it. This allows the base station controller to use the bearer channel to service multiple wireless access devices with a limited number of bearer channels.

24.3   The Wireless Data Air Interface

24.3.1   Quality

The greatest difference between landline-based data transmission and mobile wireless data transmission is the quality of the connection to the user's device. In landline communications this is generally achieved by using a twisted pair of copper wires, which has both the advantage and disadvantage of being physically fixed between two points—an advantage because the communications path is of known, fixed quality. It is a disadvantage because in today's highly mobile work environment, it is very inflexible. The wireless air link, however, is just the opposite. The air link is very flexible but has quality-of-service problems that are inherent with all radio-frequency systems. The characteristics of a wireless radio channel that affect quality include signal attenuation (free-space loss) and a very noisy environment that can cause a significantly higher bit error rate (BER) than a landline call. Also, since the data that is transmitted over this medium is digital (bits), this causes grave concern for data integrity.

24.3.2   Signal Attenuation

The wireless signal falls off at approximately $1/R^4$ (the inverse of the radius to the fourth power). Simply put, if the wireless signal is 1 W at 1 mi from a cell site, then at 2 mi it will be $1/2^4$ W or 1/16 W. This is considered to be a very rapid rate of decay (or slope of decay) of a signal. Additionally, wireless signals suffer a wide range of attenuation within buildings, ranging from 30 to 40 dB. Finally, the composition and density of objects blocking the source of the transmission from the cell site to the mobile data terminal are extremely variable by location, time of day, season, and many other factors.

24.3.3 Noise Environment

The air link is also plagued with multiple sources of interference and noise including man-made and white gaussian noise as well as effects of multipath (Rayleigh) fading. The effects of Rayleigh fading are caused by low-level refractivity associated with loss of signal propagation that is especially sensitive to variations in weather conditions and terrain. As we have seen before (Chapter 6, "Radio-Frequency Propagation"), this can be caused by flat ground, bodies of water, high-pressure barometric areas, and broad open country.

24.4 Short Message Services

The first wireless data services were short message services. Primarily used for paging, these services were typically limited to one frequency, had a low data rate, and no error correction mechanism. The pioneer networks (ARDIS and RAM Mobile Data) are still in operation today, though many technological changes have been made to improve throughput.

24.4.1 ARDIS

The ARDIS system started as a joint venture between IBM and Motorola. Initially developed in 1990, its purpose was to provide connectivity for the Motorola data radio network and the IBM field service network. Motorola later bought out IBM's interest in ARDIS. Unlike cellular systems, the ARDIS system reuses a single frequency of 25 kHz nationwide. There are up to five additional frequencies in use in selected cities where ARDIS is licensed. ARDIS uses the Motorola MDC 4800 and RDLAP protocols.

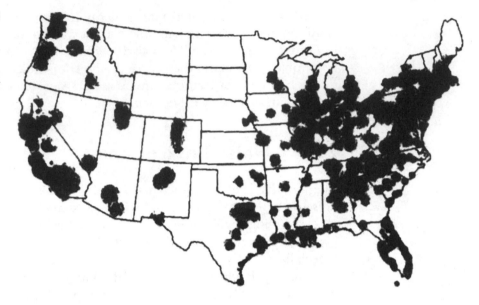

Figure 24-2
RAM Mobile Data service area. (Source: www.bellsouth.com.)

24.4.2 RAM Mobile Data

The RAM network was initially built by a consortium led by Bell-South. Consisting of over 2400 base stations configured in a cellular layout, the network uses twelve 5-kHz specialized mobile radio (SMR) channels with dozens of channel cells in each metropolitan area. Each channel provides 8 kbps throughput. See Figure 24-2 for a depiction of RAM system coverage in the United States.

24.5 Amateur Packet Radio

While the ARDIS and RAM networks were being built, the amateur radio community was also experimenting with wireless data technologies. Founded in 1983, the Tucson Amateur Packet Radio Corporation (TAPR) was one of the first groups of ham radio enthusiasts organized to work on methods for transmitting data over ham radio frequencies. In November of 1983, TAPR released its first terminal node controller (TNC). When connected to a ham radio, this device allowed for bursts of packet data to be transmitted. Figure 24-3 shows a typical packet radio network.

The TNC air interface utilized a 4800-bps half-duplex modem connected to the microphone and earphone jacks on a ham radio. Normally,

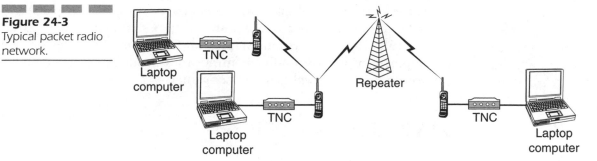

Figure 24-3
Typical packet radio
network.

the radio would be receiving all the communications on a radio channel specified by the Amateur Packet Radio (AMPR) coordinating body. When a TNC had a packet to send, it would activate the radio's transmitter and send the data.

24.5.1   AX.25

While the members of TAPR would have preferred using standard X.25, a number of additional requirements existed which forced deviation from traditional X.25, leading to the development of Amateur X.25, abbreviated as "AX.25":

1. Normal X.25 uses point-to-point leased lines for "neighbor" communications. However, as a broadcast medium, radio does not allow for the separation of conversation that is implicit with a dedicated leased line. As a result, Amateur X.25 took advantage of the "multidrop" mode of service available with X.25. In this mode the address field of the X.25 packet is used to identify the destination node of the packet.

2. FCC rules require the source of all ham radio communications to be identified by the call sign of the transmitter. To handle this requirement, TAPR changed the definition of the X.25 address field to support ASCII transmission of both the source call sign and the destination call sign.

3. It is possible for two stations to not be able to communicate with each other due to signal attenuation. However, it would be desirable for communication to be relayed by a third station that could communicate with the source and destination stations. To support such relaying, the address field was again extended—this time allowing for a list of call signs. This list specifies the stations and in what order they are to be used for relaying packets to the destination. By specifying a source route, end-to-end communications would be possible.

With these changes, multiple AX.25 users could share the "airspace" and communicate with other AX.25 systems. In a short amount of time, AX.25 bulletin board systems started springing up, allowing for the development of wireless on-line communities, including electronic mail, discussion forums, as well as multiuser games.

24.5.2   Amateur TCP/IP

Since a large amount of the packet radio community had ties to the computer science research community, many were aware of the Internet. Realizing that standards existed for the transmission of Transmission Control Protocol/Internet Protocol (TCP/IP) over X.25, a number of amateur radio enthusiasts started work on a method for running TCP/IP over AX.25. In the mid-1980s, a package was released that ran on the IBM PC allowing for the construction of radio IP networks. Called "KA9Q NET," this package became the basis for research into delivering TCP/IP over low-speed radio links.

Key: The development of the Tuscon Amateur Packet Radio Corporation (TAPR) was a very significant milestone in wireless data communications, because it became the first true instance of the use of TCP/IP in a wireless environment. This is what ultimately led to the development of other major wireless data protocols and systems, such as CDPD, WAP (Wireless Application Protocol), and other "mobile web" products. It set the stage for the marrying of the two hottest technologies of our time: wireless communications and data communications (which was driven by the Internet explosion that began in 1995).

24.6   Cellular Data Services

24.6.1   Cellular Digital Packet Data (CDPD)

The Cellular Digital Packet Data (CDPD) access protocol was developed jointly in the early 1990s by Ameritech Mobile Communications, Bell Atlantic Mobile Systems, GTE Mobilenet, Contel Cellular, McCaw Cellular Communications, NYNEX Mobile Communications, AirTouch Cellular, and Southwestern Bell Mobile Systems to provide data services on an existing analog cellular infrastructure. The CDPD protocol was standardized in the 1995–1996 time frame. These companies designed the network to support the following goals:

- Seamless service for users in all network systems
- Future growth and scalability based on standard open network protocols, including OSI Connectionless Network Protocol (CLNP) and TCP/IP
- Maximum use of available network infrastructure
- Protection of data and user's identity
- Minimal configuration parameters

Of all cellular wireless data transport systems, CDPD is probably the widest deployed. A deployment map is shown in Figure 24-4.

The CDPD architecture extends to the OSI Network Layer 3 (Figure 24-5). Mobile end systems (M-ESs) support user application data recovery in the higher layers of the OSI model. This functionality is also common elsewhere in data communications. Within a CDPD network, there are a number of network components, and the interrelationship of these systems is shown in Figure 24-6.

CDPD end systems (ESs) are the logical endpoints of communications and are the entities addressed as source or destination (service access points). Mobile end systems generally take the form of laptops or palm-

Figure 24-4
CDPD coverage area 2000. (Source: www.wirelessdata. org.)

☐ <50% coverage
■ >50% coverage
▨ Trial market

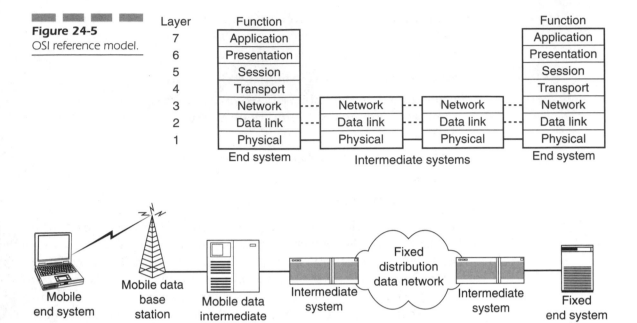

Figure 24-5
OSI reference model.

Figure 24-6 CDPD network component relationship.

tops with CDPD modems. Fixed end systems (F-ESs) host data applications on server or mainframe systems and are completely isolated from the mobility issues of the M-ESs they communicate with during transmissions. Intermediate systems (ISs) are network entities providing network relay functions by receiving data from one correspondent network entity and forwarding it to another. Intermediate systems can be physically located anywhere in the wireless system: at the MSC (MTS) or at the base station. However, it is critical that they be situated "behind" the mobile data intermediate system (MD-IS) in the network hierarchy. ISs are commonly referred to as *routers* and are unaware of mobility. Mobile data intermediate systems are the intermediate systems which provide mobility management for the M-ESs in a given service area by defining the home and service (location) functions which can span different carrier's CDPD networks. Mobile data base stations (MD-BSs) are the cellular radios designed specifically for CDPD connections.

24.6.1.1 CDPD Channel Hopping
The great efficiency available with CDPD is made possible by employing a technology known as *chan-*

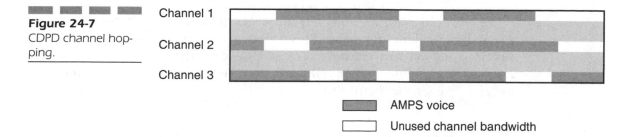

Figure 24-7
CDPD channel hopping.

nel hopping to minimize the impact on wireless carrier's facilities between cell sites and switching systems. Channel hopping utilizes the empty spaces (i.e., silence) in voice conversation to carry one or more datagrams (Figure 24-7).

Packetized data can efficiently be placed in these unused portions of the voice channel to optimize the use of available bandwidth. To do this effectively requires that the data transmission never interferes with the voice call. The MD-BS (radio) has a contention method that allows the packetized data to be applied to the unused portions of the voice channel in either a planned hop or forced hop mode.

A *planned hop* involves the "auto-sensing" of unused bandwidth and insertion of the data packet in a continuous flow without buffering. A planned hop means that there is bandwidth available on a given RF channel, but another channel has *more* unused bandwidth, so the system (the CDPD protocol) plans to move, or "hop," the data transmission to another channel to make use of even more unused bandwidth.

A forced hop means that CDPD transmission is occurring within the unused bandwidth of a channel that's carrying a voice call. Since voice traffic is considered higher in priority [quasi QoS (quality of service)], if no channels are available for an incoming voice call, the new voice call will come into the channel and kick the CDPD transmission to another RF channel.

A fixed data communication channel can support multiple connections running at 19.2 kbps. It should be noted that actual throughput after subtracting the protocol overhead is closer to 10 to 12 kbps. Also, throughput efficiency is affected by distribution of the traffic load over the entire network, the mobility of that traffic, the traffic type, and the quality of the air link. For these reasons CDPD is commonly used to support query/response, small batch entry, and telemetry applications.

Key: Forced channel hops require that the data be buffered until unused bandwidth is available. This action could cause a loss of the connection, and for this reason most carriers—over time—chose to avoid channel hopping technology altogether in their CDPD networks, and assign a fixed, static channel to CDPD traffic.

CDPD systems still exist today, but in the long term they will be replaced by a variety of web access technologies and protocols, such as WAP, running over digital wireless systems. It is important to remember that CDPD is effectively a legacy technology that was born of analog cellular systems.

24.7   Cellular Telemetry Systems

In the process of offering "mature" network services such as automatic call delivery (ACD), which was facilitated by SS7/ANSI-41 message delivery, offerings of wireless telemetry surfaced, using the SS7/ANSI-41 network infrastructure in conjunction with special modems. This service was developed and patented by BellSouth Mobility, under the trademark name Cellemetry.

Key: Cellemetry was developed when RF engineers determined that a substantial amount of the control channel's bandwidth went unused.

A number of BellSouth Mobility RF engineers realized that cellular telephone systems were capable of offering wireless telemetry services, and they developed special modems that could send messages in the control channel back to the MSC. Once at the MSC, the messages were then encapsulated in SS7 messages, and transmitted to a centralized computing platform for further processing.

Two nationwide systems exist today that implement Cellemetry. Providers of the Cellemetry service also provide the end devices (modems) and the databases, while relying on wireless carriers for the infrastructure to support the transport of the data on the control channel. The providers contract with companies to provide the services, and then pass on a portion of the revenue generated per message to the

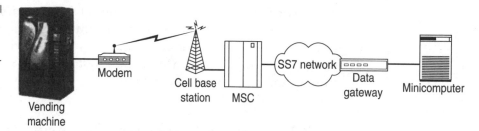

Figure 24-8
Telemetry data application.

wireless carriers. These systems allow for small amounts of data to be sent at timed intervals, or on a query/response basis, from a fixed location or a moving device. They have been successfully deployed for meter reading, copier monitoring, wireless security systems, and vending machine monitoring. Locator services use this system to transmit Global Positioning System (GPS) information on a vehicle's position. The data collected from the monitored systems is stored in a centralized database for dial-in retrieval by the company that originally contracted for the service. See Figure 24-8 for a depiction of telemetry technology.

As discussed in Chapter 18, "Wireless Interconnection to the Public Switched Telephone Network," Signaling System 7 (SS7) is used by wireless carriers to transport the ANSI-41 mobility management protocol. SS7 carries data traffic on the control channel in telemetry systems. A single wireless control channel can support 57,600 one-word messages during the busy hour! The SS7 network is engineered so that any one network link will not exceed 40% utilization under normal conditions. If a link is ever found to exceed 40%, it will be expanded. Through this design, if a link should fail, the remaining link would never exceed 80% utilization.

Key: The short bursty traffic generated by telemetry systems is of little concern to wireless network traffic engineers, since it is just a messaging service running over underutilized network resources, and it is not switched traffic.

24.8 General Packet Radio Service (GPRS)

With the advent of digital wireless technology, it became unnecessary to have a different "air interface" for data services and voice services, as voice services are now transmitted as digital information. This evolution reflects the fact that the wireless community is not immune to the industrywide efforts of convergence of voice and data traffic. Initially designed to work with GSM cellular systems, GPRS takes advantage of this change in transport technology.

24.8.1 Use with Second-Generation GSM/TDMA Systems

In a GSM or TDMA system, a wireless channel carries four interleaved sessions. Each wireless access device that has been assigned to the wireless channel knows which subchannel it has been assigned to, and will insert a burst of power into the channel when it is its turn to transmit. Since the wireless channel has a data rate of 64 kbps, the time division multiplexed subchannel has an effective rate of 16 kbps. These channels are then decompressed at the base station into four separate 64-kbps sessions and forwarded to the MSC via a fixed network. The interrelationship of these devices is shown in Figure 24-9.

However, in order to provide data services using the same air interface, a data-compatible fixed network infrastructure needs to be in place. GPRS defines this infrastructure. Figure 24-10 shows a typical GPRS network. In the GPRS network, the same GSM base station controller that is used for voice is used for data, but an additional function is required to handle the bursty nature of packet data traffic. The packet control unit (PCU) is responsible for allocating the TDMA/GSM time slots to a GPRS call. Similar to the channel hopping algorithms in CDPD, the PCU will utilize TDMA time slots in many different subchannels to support a GPRS session. As a side effect, it is possible for a GPRS session to achieve a maximum throughput of 64 kbps! An example of GPRS time slot allocation is shown in Figure 24-11. The PCU will in turn send the GPRS

Figure 24-9
GSM/TDMA wireless network.

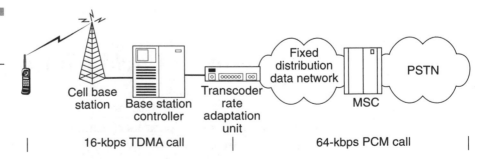

Cell base station   Base station controller   Transcoder rate adaptation unit   Fixed distribution data network   MSC   PSTN

16-kbps TDMA call   64-kbps PCM call

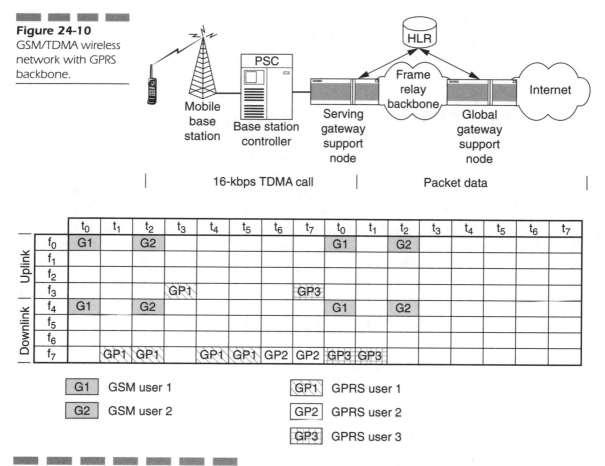

Figure 24-10
GSM/TDMA wireless network with GPRS backbone.

		t_0	t_1	t_2	t_3	t_4	t_5	t_6	t_7	t_0	t_1	t_2	t_3	t_4	t_5	t_6	t_7
Uplink	f_0	G1		G2						G1		G2					
	f_1																
	f_2																
	f_3				GP1				GP3								
Downlink	f_4	G1		G2						G1		G2					
	f_5																
	f_6																
	f_7		GP1	GP1		GP1	GP1	GP2	GP2	GP3	GP3						

G1	GSM user 1
G2	GSM user 2

GP1	GPRS user 1
GP2	GPRS user 2
GP3	GPRS user 3

Figure 24-11 GSM/TDMA time slot use.

packets to the serving gateway support node (SGSN). This is effectively a router, taking the V.110 data stream from the end system, and placing it into a frame relay PVC. Like the MTS, the service gateway support node is responsible for managing the handoff of GPRS sessions from one base station controller to another, and therefore will interface with the HLR.

At the edge of the GPRS backbone is the global gateway support node. This device acts as a firewall, protecting the GPRS infrastructure from any attack that may come from the Internet. Further, this device also participates in the HLR operations, so it can route incoming traffic to the appropriate SGSN.

Author's note: News reports in industry trade magazines circa October 2000 stated that the GPRS technology was a stop-gap measure, an evolutionary step to higher-level wireless data transmission strategies. It is predicted that the GPRS technology would "die" as it is known today, by the year 2002.

24.9   EDGE (2.5 Generation)

Initial deployments of TDMA and GSM utilized GSMK for encoding digital data onto the air interface. However, with the improvement of digital signal processor (DSP) technology, it is now possible to use 8 PSK (phase shift key). This allows for 3 bits to be encoded per analog symbol, causing the TDMA subchannel to become a 48-kbps data bearer. When coupled with time slot hopping, a GPRS session is now able to achieve a 192-kbps data rate.

24.10   Wireless LANs

In all the previous sections, we have discussed wireless data technologies that allow for full user mobility. While these technologies allow for the greatest amount of mobility, it comes at a cost: lower data rates. As a result, wireless LAN technology has become quite popular for installations that do not require the same level of mobility.

24.10.1   Wireless Ethernet

The wireless Ethernet standard is a product of the IEEE Ethernet Working Group, 802.11. Utilizing channels in the 2.4-GHz industrial, scientific, and medical (ISM) band, this standard allows for 2-Mbps and 11-Mbps data rates at distances up to 3.5 mi. Work on 802.11 continues with efforts focusing on quality of service (QoS) and development of increased rates, up to 20 Mbps.

24.10.1.1   Air Interface   The 802.11 standard allows for two different radio interfaces: direct sequence spread spectrum (DSSS) and frequency hopping spread spectrum (FHSS). As their names imply, these radio interfaces really only differ in how the ISM-band frequencies are used to carry the wireless Ethernet data. Both approaches are derivatives of spread spectrum, a modulation technique that spreads the transmitted signal over a frequency range much wider than the minimum bandwidth required to send the signal. While using a larger bandwidth footprint increases the possibility of interference, the actual amount of interference is decreased because the data ends up being replicated across the frequency spectrum. When a transmitter and receiver use the same base pseudorandom signal to "tune" in the same signal, interference is effectively canceled out. This allows for higher reliability at signal levels below the ambient noise floor.

Direct sequence spread spectrum, probably the most widely utilized form of spread spectrum, modulates the digital data onto a carrier wave, and then remodulates the resulting signal onto a pseudonoise signal, similar to CDMA transmissions. This causes the spectrum used for one burst of power to be significantly widened.

Frequency hopping spread spectrum works by modulating the digital data onto different frequencies based on a pseudorandom noise signal. This causes the frequency in use to instantaneously jump based on the pseudorandom input. As a result, the frequencies used are spread across the entire spectrum. In addition, multiple frequencies may have the same information signal transmitted at the same time. The number of redundant frequencies in use in part determines the immunity realized by this modulation scheme.

24.10.1.2 Operational Considerations—Wireless LANS The 802.11b wireless network can be operated as either an ad hoc network of neighboring workstations or as an "infrastructure network" connecting to a wired Ethernet backbone. When a number of access points from the same vendor are deployed, they may be able to coordinate session handoff, similar to cellular and PCS systems. However, intervendor handoff currently is not possible, as the 802.11b standard does not specify a signaling method to coordinate the handoff. Vendors have thus ended up using different and/or proprietary signaling methods for the time being. Work on a standard inter-access point protocol is under way. See Figure 24-12 for a depiction of a typical 802.11b network topology.

Corporations deploying wireless Ethernet in either ad hoc or infrastructure network modes should also be concerned about unauthorized computers participating in their wireless Ethernet networks. Today, the 802.11b standard only provides for a clear-text "network ID" to determine whether a node is allowed to participate in a network. As a result, many vendors are now adding 40-bit and 128-bit public-key encryption to their 802.11b network devices.

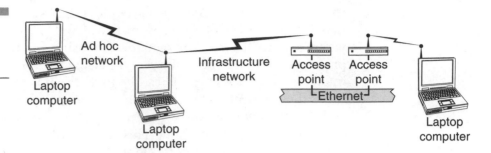

Figure 24-12
Wireless Ethernet
(802.11b) network
topology.

24.10.2   Bluetooth

Similar to 802.11b, Bluetooth also uses the 2.4-GHz ISM band to form ad hoc interconnections between computing devices. Initially designed as a method for eliminating the interconnect cables needed to connect laptop, cell phone, and personal digital assistant, Bluetooth is also being proposed as a method for e-commerce and local voice communications.

24.10.2.1   Wireless Interconnect   As shown in Figure 24-13, Bluetooth allows all the personal electronics used by an individual to communicate with each other. This allows for a number of new applications, including automatic and dynamic synchronization of information stored in a person's pentop, laptop computers, wireless phone, and company LAN. Furthermore, devices like wireless phones can become "gateways" from the Bluetooth LAN into switched or packet radio networks.

24.10.2.2   Local Voice   The Bluetooth channel format also includes three 56-kbps voice channels. These are intended to allow people with wireless phones to directly communicate with other wireless phone users who are in close proximity, as well as with the PSTN, through home-deployed base stations. The addition of these usage modes to mobile phone technology will help solidify wireless phones as the one-stop communication device of the future. These scenarios are shown in Figure 24-14.

24.10.2.3   E-Commerce Platform   Bluetooth has also been proposed as a method for transmitting e-wallet information to a point of sale (PoS) terminal. The point of sale terminal will then authenticate the e-wallet information, and charge the account referenced by the e-wallet for the items purchased. These accounts could be bank accounts, credit card accounts, or even mobile service provider accounts. Target applications include smart gas pumps, toll highway collection systems, and vending machines. See Figure 24-15 for a depiction of a typical Bluetooth e-commerce architecture.

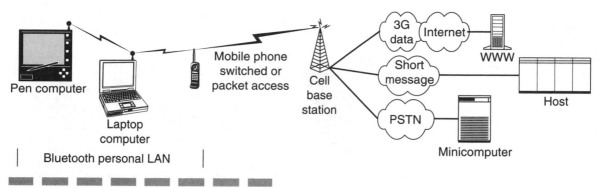

Figure 24-13 Typical Bluetooth "personal LAN."

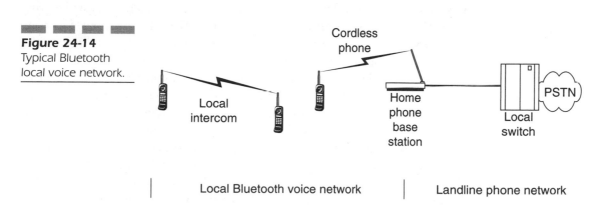

Figure 24-14
Typical Bluetooth
local voice network.

24.10.2.4 Operational Considerations Bluetooth operates currently within a range of only 30 ft, much lower than the 3.5 mi of 802.11b. This is a direct consequence of Bluetooth's goal to place all the electronics necessary to make a transceiver into one integrated circuit (IC). Work is under way to extend this range to about 300 ft. But given the "personal LAN" concept, limited range also provides some implicit, built-in security.

24.11 Application Protocols

24.11.1 Wireless Application Protocol (WAP)

While all the protocols described heretofore allow for greater mobility, they all provide relatively low data rates. New application protocols that take the data rate into account are therefore desirable. The first such protocol is the Wireless Application Protocol (WAP). The WAP Forum was

Figure 24-15
Typical Bluetooth e-
commerce solution.

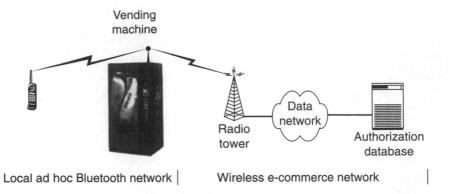

Figure 24-15
Typical Bluetooth e-commerce solution.

developed jointly by Ericsson, Nokia, Motorola, and Phone.com in June 1997. The objective of the founders of the WAP Forum was to create standards allowing small, consumer-class wireless devices to access the Internet anywhere, anytime.

WAP devices are limited in the sense that they don't have the luxury of accommodating large code space to run applications, unwieldy kernels, and expansive displays. Developing applications "in the small" is given new meaning on a WAP device.

In the WAP environment, wireless devices do not need to have a complete web browser implementation to provide web access. Instead, the wireless service provider will provide WAP protocol gateways to reduce the amount of data that needs to be sent to the wireless device, to the bare essence of a web page in terms of the total number of bytes of data required. This is accomplished by looking for specific wireless markup language (WML or Wireless XML) tags embedded in the web page being accessed. As a result, the transmission of high-resolution graphics, wordy text descriptions, and objects with complex formatting can be skipped in the transmission through the wireless carrier's infrastructure to the wireless handset.

Once reduced, the web page is then sent to the wireless device using a lightweight transport stack based on User Datagram Protocol (UDP) and compressed IP. Use of compression is desirable as it allows the bearer channel to be used as efficiently as possible. Instead of a typical 64-byte TCP packet containing 15 bytes of payload, the compressed packet will require approximately 20 bytes of data total, to be transported to the mobile handset.

24.11.1.1 WAP Subprotocols Bandwidth conservation is crucial in the wireless world. WAP provides a protocol stack designed to minimize bandwidth requirements while offering secure connections over a variety of transport protocols. The WAP architecture is straightforward and looks similar to the pervasive ISO network models. These subprotocols and their functions are listed below:

■ The *WAP application environment* (WAE) is the topmost layer and the one that application developers are most interested in. This layer holds general device specifications, WML and WML Script programming languages, the WTA telephony application programming interfaces, and defines content formats, including graphics and personal information management (PIM) information.

■ The *Wireless Session Protocol* (WSP) in the session layer is a tokenized version of HTTP 1.1, designed specifically for limited bandwidth and longer latency applications while still allowing guaranteed delivery and "push" content.

■ The transaction layer encompasses the *Wireless Transaction Protocol* (WTP), which limits the overhead of packet sequencing. It supports a variety of message types, including various types of guaranteed and nonguaranteed one- and two-way requests.

■ The security layer supports *wireless transport layer security* (WTLS), a protocol based on standard TLS, formerly known as secure sockets.

■ WAP's lowest level contains the *Wireless Datagram Protocol* (WDP), which provides a consistent interface between various over-the-air protocols, including CDMA, CDPD, GSM, TDMA, and ReFlex.

The interrelationship of these protocols, and more traditional Internet protocols are shown in Figure 24-16.

24.12   Wireless Data Technologies—Summary

Just as the Internet considerably transformed the wired communications market from 1995 to 2000, the next 4 years (2001 to 2005) will forever change wireless communications. New products, technologies, and applications are beginning to be deployed with an evolution from a voice focus to a data-centric multimedia experience. In 2001 and 2002, new air interfaces will be deployed, significantly increasing data rates. However, the benefit of the wireless Internet is based on its mobility, its always-on nature, and convenience, not necessarily its bandwidth. It is estimated that the number of worldwide wireless data subscribers will grow from more than 133 million to more than 620 million by the end of 2004.

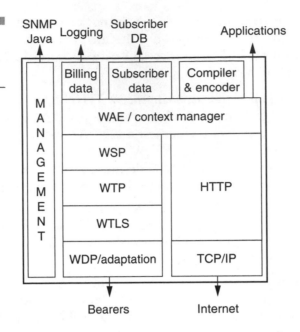

Figure 24-16
Wireless Application
Protocol gateway
architecture.

24.13   Test Questions

True or False?

1. _____ In a typical cellular system, a significant amount of the control channel bandwidth is unused.

2. _____ IS-95 and IS-136 are the protocols that are used to carry data traffic on the control channel in telemetry systems.

3. _____ In the realm of mobile wireless data, the two most common forms of data transmission are circuit-switched (portable) communications and packet data (fully mobile) communications.

Multiple Choice

4. Today, one of the most prevalent wireless packet protocols is:
   (a) RAM
   (b) Cellular Digital Packet Data (CDPD)
   (c) ARDIS

(d) AMPS

(e) None of the above

5. The predominant types or applications of wireless data include:
   (a) Bursty communications (i.e., file transfers)
   (b) Terminal server sessions
   (c) Query/response applications
   (d) Batch file applications
   (e) Web server access
   (f) a, c, and d only
   (g) a, b, and d only
   (f) e only

6. The great efficiency available with CDPD derives from a technology known as:
   (a) Channel stuffing
   (b) Noncontinuous transmission
   (c) Channel hopping
   (d) W-CDMA

7. What is the name of the protocol that describes the effort to develop a wireless LAN standard?
   (a) ANSI-41D
   (b) 802.3
   (c) 802.11b
   (d) 802.5
   (e) None of the above.

Short Answer/Essay

8. Name and describe the function of each of the subprotocols of WAP.

9. Describe the multiple functions of the Bluetooth initiative as they relate to standard wireless LANs.

CHAPTER 25

Commercial and Business Issues

25.1 CTIA and PCIA

There are two distinct but complementary industry trade and advocacy groups in the United States that support the wireless industry: the Cellular Telecommunications Industry Association (CTIA) and the Personal Communications Industry Association (PCIA). Both groups are head-quartered in Washington, D.C. The CTIA supports cellular carriers nationwide, while the PCIA supports PCS carriers nationwide.

Wireless carriers must become member companies of these organizations in order to reap the benefits offered by these trade groups. Membership benefits include regular access to industry reports regarding business and technology issues, industry statistics, and contact information for other carriers that lists persons by title and department. The CTIA and PCIA are also involved in representing the wireless industry to the FCC and take part in legislation that will impact the industry by attending hearings and testifying before Congress regarding issues and/or laws that will impact the industry.

25.2 Wireless Industry Growth Statistics

25.2.1 1983–1994

Wireless telephone service is the fastest-growing consumer technology in history, even surpassing the VCR's growth in the 1980s. From 1983 to 1995, the wireless industry created more than 40,000 new jobs and invested $16 billion in related equipment and facilities. The cellular industry shattered records as it added 3.3 million subscribers—the largest half-year increase ever—in the first 6 months of 1994. The industry set its twentieth consecutive record for 6-month revenues by July 1994, reaching $6.5 billion, a 35% increase over 1993's first-half revenues.

As of June 1995, there were approximately 27 million cellular telephone customers in the United States. From July 1993 to July 1994 the industry picked up more than 6 million new subscribers, a growth rate of 48%. The annual growth rate for most cellular companies in the 1980s was anywhere from 45 to 60%. Two out of every three new phone numbers are assigned to wireless customers. This is one reason for the large number of new area codes being implemented nationwide. Cellular service took 9 years to reach 10 million subscribers, but in less than 2 years (October 1992 to June 1994) the number of cellular customers doubled. The astounding growth in wireless service is ascribed to the public's increasing interest in the convenience, safety, and work productivity offered by wireless phones.

25.2.2 1997

The wireless industry in the United States gained 11.3 million subscribers during 1997, according to a survey by the CTIA. This includes the following industries: cellular, PCS, and enhanced specialized mobile radio (E/SMR). The industry also saw its annual revenue jump by 16.3%, from $23.6 billion in 1996 to $27.5 billion in 1997, while the average monthly bill declined from $47.70 to $42.78 (see Section 25.3, "Monthly Wireless Phone Bills Decreasing"). Still, CTIA vice-president Tim Ayers points out that the lack of "calling-party pays" service in the United States has hampered subscriber growth and wireless phone usage (see Section 25.9, "Prepayment Offerings").

25.2.3 2000

Table 25-1 shows the astounding growth in the wireless industry in the United States from the earliest days of the industry (1985) to 1999. These numbers reflect information from both cellular and PCS carriers, and show the economic and cultural impact that this industry has had in America since the mid-1980s. Table 25-2 shows additional growth and various industry statistics as of December 1999.

Key: In the year 2000, the CTIA estimates that an average of 46,000 people signed up for wireless service in the United States each day. Experts estimate that by 2005, there will be over 1.26 *billion* wireless phone users around the world.

25.3 Monthly Wireless Phone Bills Decreasing

According to industry statistics, as cited in Table 25-1, the average monthly wireless bill dropped from $98.02 to $41.24 between 1985 and 1999.

TABLE 25-1

Industry Growth
1985–1999

Wireless Industry	1985	1999
Total subscribers	91,600	103,000,000*
Total revenues	$482,000,000	$40 billion
Average monthly local bill	$98.02 (1988)	$41.24
Industry employees	1404	155,817
Average length of local call in minutes	2.33 (1987)	2.38
Cell sites in service	346	81,698
Cumulative capital investment	$911 million	$71.3 billion

*This number reflects total subscribers in the United States as of October 2000.

TABLE 25-2

Wireless Industry
Growth Facts, circa
2000

Number	Factoid
385 million	Worldwide wireless subscribers
3 million	Wireless Internet subscribers
150 billion	U.S. wireless minutes of use (MoU)
20%	Increase in wireless MoU each year
32 cents per minute	Average price per minute for U.S. wireless service
207 million	Handsets sold
58 million	Consumers using next-generation pagers with e-mail functionality
48%	U.S. households with wireless phones
15.7%	Kids with wireless phones
55.3%	Chicago subscriber penetration rate
54%	Charlotte, N.C., subscriber penetration rate
5%	Worldwide subscriber penetration rate
25%	Decrease in price per minute each year

That represents a *60% decrease* in the average monthly wireless bill! This drop is an important industry barometer, and results from multiple factors:

- Over the years, the competition in the wireless industry has become greater and matured. This includes the entry of wireless resellers into the marketplace, along with more aggressive marketing campaigns. In the late 1990s, the entry of PCS carriers onto the wireless landscape further escalated the competitive atmosphere in the industry. This development contributes to lower monthly bills because prices charged to consumers are dropping, ultimately in the form of lower cost per minute to place (and receive) calls.

- As the wireless industry has matured, wireless carriers have gained experience and become more adept at building and managing their networks. This ultimately has lead to more efficient fiscal responsibility and streamlined operations. Again, this lowers their costs, which can be passed on to the consumer. It should be noted that the carriers have been *forced* to find ways to lower their costs due to competition in the marketplace.

- Another reason for the lowering of monthly bills over the years is because a key element in the huge increase in subscribers in the industry has been the increase of more and more subscribers from the general population versus the business segment. More subscribers, overall, means that revenues go up, which may allow for lowering of some price elements.

- As the industry has matured, equipment suppliers have streamlined their own operations as well, producing equipment infrastructure for the industry more cheaply yet more efficiently. They have also had much higher volumes to produce, as the PCS carriers have built out their markets. The increased overall volume has probably allowed them to charge less per unit, because they can still attain specific profit targets through volume versus price per unit sold. Again, this lowers the carrier's costs, which can be passed on to the subscriber.

25.4   Cost per Gross Add

One of the financial indicators of a healthy wireless company is a *low* "cost per gross add," known as *cost of acquisition* in other industries. The cost per gross add is the gross cost to the company to obtain one new customer, in terms of the following costs and overhead: executive management, marketing programs, engineering, operations, customer service, information technology (IT) systems maintenance and development, and corporate administration. A large increase in nonbusiness customers is good only if the cost of customer acquisition (cost per gross add) is not high.

25.5   Average Revenue per Unit

Another financial indicator that is an offset to cost per gross add is average revenue per unit (ARPU). This statistic indicates how much revenue per subscriber (per customer) a wireless carrier obtains on a monthly (regular) basis. The higher the ARPU, the better. ARPU can also be analyzed in conjunction with cost per gross add, so a wireless carrier can determine what their *net* revenue is per subscriber. ARPU shows up frequently in wireless carrier financial reports.

25.6   Per-Minute Rates

The industry average cost to build an entire cell site ranges from approximately $350,000 to $750,000. This cost is lower for PCS cell sites. It includes land for the site (usually a lease), taxes, shelter, base station equipment, tower, antennas, and construction labor. The huge cost of wireless infrastructure is why customers used to pay such high rates per minute for cellular phone use, compared to *landline rates*. The cost to build, maintain, operate, and continuously expand and improve a wireless network is enormous. This is especially true in urban areas with large populations, a denser concentration of cells, and higher expectations from customers.

Key: As cellular coverage continues to expand across the United States, eventually the cost per minute charged to cellular subscribers will begin to trend downward. As coverage increases, there will be less need for new infrastructure, and therefore less capital investment required by wireless carriers. Ultimately the decrease in costs will be passed on to the consumer in the form of lower rates.

Wireless rates per minute that are charged to subscribers average around 10 to 15 cents today (circa 2000). Just 5 years ago the average cost per minute charged to wireless (cellular) subscribers was anywhere from 22 cents per minute to 35 cents per minute. As wireless coverage approaches the saturation point in the United States, there will still be a need to continually build more base stations to increase system capacity and coverage. But the *rate* at which new base stations will be required will begin to slow, driving down costs (as stated in Section 25.3, "Monthly Wireless Phone Bills Decreasing"). This evolutionary development, along with increased competition, will result in wireless service becoming more of a commodity (a more common, inexpensive product). This is actually already happening. The more this service becomes a commodity, the lower the prices charged to consumers. This is what will catalyze the push to replace wireline phones with wireless phones.

25.7 Churn

Churn is a marketing term that refers to the number of customers who cancel their wireless service. In other words, it describes the number of customers who "churn" off the system. There are two types of churn:

- Intercarrier churn occurs when customers, for whatever reason, cancel service with their existing service provider because they have obtained service from *another* service provider in their market.

- System churn occurs when customers, for whatever reason, cancel service with their existing service provider and do not obtain service from a competing wireless carrier in their market. These are customers who no longer desire *any* wireless service whatsoever.

Unfortunately for wireless carriers, customer acquisition cannot come close to matching churn rates, especially since 1997–1998, when the PCS carriers began to dig in and compete aggressively with the incumbent cellular carriers. The key to making a difference in the wireless marketplace is to provide *value* that the competition does not provide. In 1998, annual churn in the industry reached almost 26%! Four million people signed up for "new" service that year, but 16 million people switched providers or cancelled service altogether!

25.8   Agents and Resellers

Many businesses have developed over the years to act as sales agents for wireless carriers. These companies purchase huge blocks of air time from wireless carriers at a discounted, wholesale rate. They then resell the airtime to their own customers at a markup, acting as their own independent wireless company from the customer's perspective. It is easy to tell who the resellers are in any given market, because they have names that are distinct and separate from the "A" and "B" licensees.

25.9   Prepayment Offerings

Around 1995, industry studies revealed the fact that up to 25% of *potential* cellular customers fell into the category of "credit challenged." Because of this newfound information, many cellular and PCS carriers have scrambled to find ways to sign on credit-challenged customers in a manner that minimizes risk to the carriers. Almost all of these systems work as follows:

■ These customers can buy a mobile phone, and purchase blocks of minutes of use in increments of $50, $100, $150, and so on. No credit check is necessary, and that particular administrative overhead is avoided.

■ At periodic intervals the system (the wireless switch in conjunction with the carrier's billing system) notifies the customers how much airtime they have left, in dollars or minutes of use. Customers can also pull that information up on demand on their cellular or PCS phone.

Prepayment systems offer important benefits: First, wireless carriers can harness a large portion of their potential customer base essentially risk-free. Second, this system becomes an option even for potential customers who have good credit but who want to maintain strict control over their wireless expenditures.

25.10   Cellular Market Ownership

Today, cellular markets are bought and sold on a regular basis among cellular carriers nationwide. The FCC must approve all market sales.

Many times, markets are *traded* between cellular carriers for each other's cellular markets of equal or comparable value. There are advantages and disadvantages to trading markets with other cellular carriers.

25.10.1   Advantages of Buying and Trading Markets

The main advantage of trading markets is that it allows cellular carriers to build out their existing cellular *clusters.* A cellular cluster is a contiguous geographic arrangement of adjacent cellular markets, where all the markets are owned by the same company.

The main benefit of acquiring cellular markets to enlarge clusters is that it allows carriers to attain a competitive edge and to exploit economies of scale in terms of network design and configuration, marketing, and system operations.

Sometimes, cellular carriers even buy or sell just *portions* of their cellular markets, usually a county or two. There are strategic reasons for doing so. For instance, a cellular company may purchase only a *county* in an MSA or RSA because an interstate highway ("I-road") running through the county links two of the carrier's existing markets. That way, if a customer were having a conversation on her cellular phone, and crossed into the county linking the two cellular markets, the call would not transfer to the neighboring market's system and become a roaming call.

25.10.2   Disadvantages of Buying and Trading Markets

There is one main disadvantage to buying or trading for other cellular markets: The incoming cellular carrier is inheriting a *legacy* system. The previous owner of a cellular market and the employees hired to design and operate the system may not have been very knowledgeable or competent about cellular system design and operations. Therefore, there may be an extensive amount of rework involved in redesigning the cellular system where design deficiencies existed under the previous owner. This would cost a lot of labor time, and possibly a large amount of capital to invest in new equipment. For consistency's sake, the incoming carrier may have all-new equipment installed in the market, so that the equipment in all of the carrier's markets is homogenous. This would greatly facilitate network management and maintenance operations.

Another disadvantage is that the previous owner of the market might have inked bad contracts and deals with any number of other businesses:

■ There may be unfavorable interconnection arrangements with area local exchange carriers.

■ Unfavorable leases may have been signed for property, buildings, or tower occupancy. For example, the previous market owner may have signed a lease for an MSC site that is locked in for 20 years. In a rapidly evolving industry, that would be very foolish. The incoming cellular company may desire another, better location for an MSC site.

25.11   Market Values

Some cellular carriers may own just a single market in a large area, where they are surrounded by many other cellular markets owned by other carriers. Of course, one cellular market does not constitute a cellular cluster, which strategically is the ideal manner in which to build out a cellular system. In this situation, it may seem obvious that a cellular carrier should sell that single cellular market. However, in reality it may be a very wise decision for the cellular carrier to maintain ownership of that one cellular market so that its value increases to other regional carriers who may want to purchase that market. The longer that the adjacent market owners have to wait to make that single market part of *their own* cluster, the more valuable that single market will become over time. This concept applies mainly to areas where there is a dense concentration of POPs or a substantial amount of interstate highways (I-roads).

25.12   National Branding and Marketing

Across the United States, many "A" band carriers use the brand name Cellular One when selling wireless service. This marketing right dates back to the earliest days of the cellular industry. Back in the early 1980s when cellular was still an infant industry, the existing industry players thought they would be able to solidify their marketing efforts if they all offered a consistent brand name across the country. At that time, McCaw Cellular was the dominant "A" band cellular carrier in the United States. McCaw trademarked the Cellular One name brand by early 1984, and later Southwestern Bell Mobile also bought the rights to use the Cellular One brand name in all its "A" band cellular markets when it acquired the Chicago "A" band market. The right to use the Cellular One brand name comes with certain requirements, most of them centered around customer service. For example, using the Cell One brand name requires the carrier to maintain 24×7 (24 h per day, 7 days per week) customer service.

In 1984, all of the "B" band carriers were still associated with the Regional Bell Operating Companies (RBOCs). They intended to trademark and use the "Mobile One" brand name in all "B" band markets across the United States. But the "B" band carriers backed away from this initiative because of their fear that they would be charged with some type of antitrust actions because the consent decree (breakup of the AT&T/Bell monopoly) was still fresh in their minds. The Mobile One brand name was then unprotected, and it started getting used by other entities (i.e., paging carriers, cellular resellers). Since there was then no restrictions on what companies could use that brand name, its marketing effectiveness became diluted for the "B" band cellular carriers. Therefore, they ceased their efforts to use the Mobile One brand name.

25.13 Calling Party Pays (CPP)

Since the early 1990s, there have been sporadic efforts in the wireless industry to develop and deploy a feature known as "calling party pays" (CPP). This feature would entail having persons who call wireless subscribers pay for the call themselves, instead of the wireless subscriber having to incur the cost of the call. This effort never really took root in the United States for multiple reasons.

The FCC was supposed to issue an order relating to calling party pays in February 2000, but it pushed back the date. Some wireless operators and industry observers have since suggested that the prevalence of "bucket" payment plans diminishes the importance of CPP. "Bucket" plans are those plans where the consumer can purchase "500 minutes per month for $50."

Internationally, observers credit CPP—which is the norm everywhere but in the United States—with a higher ARPU and increased wireless penetration. U.S. bucket plans have produced similar results. In fact, AT&T Wireless, the creator of "Digital One Rate"—the hugely successful bucket plan—recently discontinued its CPP offering in Minneapolis, which had been available for several years.

Some CPP offering haven't fared well because some roadblocks require operators to introduce limited offerings. Most CPP services today are only available in limited markets where long distance calls can't be charged to the caller and CPP subscribers can't roam with the service. Most industry observers believe that CPP can be successful on a limited basis, but on a nationwide basis—which would be ideal—it would be incredibly difficult and not necessarily worth the effort put into it. Billing is the biggest obstacle to widespread CPP offerings. A common billing method is for wireless operators to partner with local exchange carriers (LECs) that agree to bill the landline customers that place calls to CPP subscribers. On a nationwide basis, however, that could mean partnering with thousands of carriers. In addition, some wireless operators say that some LECs have refused to bill for the service.

25.14 Cultural Backlash

While the public's affection for wireless service has been proven by the fact that over 100 million Americans now subscribe to wireless service, a movement has been brewing in the United States since around 1998. This movement is not pro-wireless. Many in our society feel that the use of mobile telephone service has become *too* ubiquitous, and some factions of society are now beginning to draw a line in the sand. This sentiment can be summed up by saying that many people in today's society feel that wireless service makes people believe "the world is their phone booth." Some examples are given here:

■ The newsletter of a commuter rail line in Chicago had a huge, highly emotional response to one letter to the editor that complained about wireless phone use on the trains. Most people claimed it was their right to use the phones on the train, as long as they weren't boorish about it (i.e., speaking too loudly).

■ In 1999, the captain of the Chicago White Sox baseball team banned wireless phone use in the clubhouse, because too many players (usually the single players) would come off the field after a game and begin calling their girlfriends. This prevented the team from beginning their daily postgame conferences. The White Sox placed a sign with the international "banned" symbol in their clubhouse. Inside the red circle with the slash across the middle was a picture of a mobile phone.

■ Many movie theaters today flash a message across the screen prior to starting the show, that not only reminds patrons of all the goodies they can buy at the concession stand, but asks them to "please silence cell phones and pagers."

■ Many local governments have outlawed wireless phone use while driving in a vehicle, or may outlaw it. For instance, Hong Kong instituted a law in 2000 that fines anyone caught using a wireless phone while driving $256. The City of Chicago has begun hearings on this same matter as of October 2000. This is one of the reasons that the earpiece option has become very popular with wireless phones. Users can talk on the phone while having their hands free to drive.

Eventually, there will probably come a time when there will be some official, and probably some unofficial, rules of etiquette that will apply to the use of wireless phones in "civilized" society.

25.15   Test Questions

Multiple Choice

1. According to CTIA estimates, how many customers docs the cellular industry add every day?
   (a) 280,000
   (b) 17,000
   (c) 57,000
   (d) 46,000
   (e) None of the above

2. Which acronym below denotes the statistic that indicates how much revenue per customer a wireless carrier obtains on a monthly (regular) basis?
   (a) CPU
   (b) RAM
   (c) ARPU
   (d) CFS
   (e) None of the above

True or False?

3. _____ One of the financial indicators of a healthy cellular company is a high cost per gross add.

4. _____ The main advantage of buying/trading for new cellular markets is that the incoming cellular carrier is inheriting a legacy system.

5. _____ *System churn* describes when cellular customers cancel service with their cellular provider, and obtain service from the competing, opposite band cellular carrier in their market.

6. _____ Around 1995, industry studies revealed the fact that up to 25% of potential cellular customers fell into the category of "credit-challenged."

Short Answer/Essay

7. List and briefly explain all the issues behind the decrease in monthly wireless bills.

CHAPTER 26

Enhanced Specialized Mobile Radio

26.1   SMR Overview

Specialized mobile radio (SMR) service was created by the FCC in 1974 for carriers to provide two-way radio dispatching to the public safety, construction, and transportation industries. SMR systems provide dispatch services with push-to-talk technology for companies with multiple vehicles. These services included voice dispatch, data broadcast, and mobile telephone service, but SMR had limited roaming capabilities. As mentioned in Chapter 24 ("Wireless Data Technologies"), RAM and ARDIS are wireless data services that are licensed in the SMR frequency range. An SMR subscriber could interconnect with the PSTN much like a cellular subscriber.

SMR systems traditionally used one large high-power transmitter to cover a wide geographic area, like MTS/IMTS systems. This limited the number of subscribers because only one subscriber could talk on one frequency (channel) at any given moment.

Key: SMR systems operate in the following frequency ranges: 806–821 MHz and 851–866 MHz.

The number of frequencies allocated to SMR is smaller than for cellular and PCS, and there have been multiple operators in each market in the past. Because dispatch messages are short by nature, an SMR system can handle many more dispatch-only customers than interconnect subscribers.

26.2   Migration to Enhanced Specialized Mobile Radio: Nextel Corp.

In 1987, a newly founded company, Nextel Corp., began to revolutionize the SMR market. Formerly known as Fleet Call, Nextel acquired radio spectrum in six of the largest SMR markets in the United States: Los Angeles, New York, Chicago, San Francisco, Dallas, and Houston. In the early 1990s, Nextel accumulated more SMR licenses, allowing it to piece together a nationwide footprint of sorts.

In April 1990, Nextel asked the FCC for permission to build *enhanced* SMR systems in those six markets. Those new systems would be known as enhanced SMR (E/SMR). The E/SMR systems would consist of lower-power, multiple transmitters which would allow the same frequencies to be reused some distance away, very similar to cellular and PCS. The resulting cellular-like network would open new consumer and business communications markets to the SMR industry. Nextel also planned on implementing its new systems as all-digital (TDMA) systems to further expand calling capacity. In February 1991, the FCC approved Nextel's request, and its first digital mobile network came on-line in Los Angeles in August 1993.

Nextel had respectable growth in the late 1990s, presenting a challenge to the traditional wireless world, the cellular and PCS carriers. At various times, Nextel has made significant leaps ahead of its wireless competitors in service development, marketing strategy, and most recently, wireless data. Nextel's former CEO Dan Ackerson was quoted in August 1999 as saying, "We've gone from being a niche competitor that a lot of carriers dismissed, to someone that a lot of carriers fear."

Nextel is open about its market strategy of pursuing the business segment versus the consumer segment. One of the features it touts is the always-on capability that allows the Nextel phone to act as a walkie-talkie. Additionally, in conjunction with the feature described above, Nextel touts its ability to pool groups of subscribers together to have what amounts to anytime capability to have conference calls with a pre-subscribed group or distribution list. Because of these features, landscaping and construction companies are especially good fits for Nextel service and its unique features.

MCI flirted with buying Nextel at one time, but the deal was called off because Ackerson determined that the merged entity would benefit MCI much more than Nextel. The buyout was called off for that reason, and because, as Ackerson later said, "We never had a 'For Sale' sign out, and we don't have one out now." In February 1999, at the CTIA's annual trade show, Nextel made news when it unveiled plans for a wireless Internet portal. At the time, the alliance partners for what would become Nextel Online included Motorola, Unwired Planet (now phone.com), and Netscape Communications.

In mid-1999, Microsoft announced a $600,000,000 investment in Nextel, and Nextel said it would provide its Nextel Online customers with access to wireless Internet services through a customized version of the Microsoft Network portal. Netscape was quietly removed from the plan. Supposedly, there were technology, business, and financial considerations for Netscape's ouster from the alliance. Nextel's choice of portal partner for its wireless Internet venture is important, but the mere fact that the carrier is making a strong entry into advanced data service is significant on its own. At a time when Nextel's voice services have already contributed to a successful business, it is now adding a component that most other wireless carriers still are developing. The only technology that comes close is cellular digital packet data (CDPD), which has found niche success but never emerged as a mainstream wireless Internet platform.

With the rollout of its data approach, Nextel set one more milestone for other wireless service providers to meet. The company was among the first to pursue the buildout of a national network and bundle long distance charges into its service package. It was the first to eliminate roaming charges and implement per-second billing. It remains the only carrier with a national footprint to offer both a dispatch function and a conventional digital wireless voice. As wireless network technology and service quality have leveled off in recent years, wireless service providers taking aim at individual consumers and business customers are reduced to competing primarily on network reach and service price. Nextel has made a great effort to avoid becoming engaged in a price war on that front, choosing instead to focus on functionality and business value.

Nextel has clearly made great strides since its beginnings as a wireless dispatch operator saddled with technological problems and facing capable and competitive companies in the crowded cellular and PCS fields.

26.3   Test Questions

Multiple Choice

1. Specialized mobile radio (SMR) service was created by the FCC in 1974 for carriers to provide what type of service?
   (a) Wireless LAN systems
   (b) Wireless PBX services
   (c) Two-way radio dispatch
   (d) CB radio

2. Which company asked the FCC for permission to develop and build enhanced SMR (specialized mobile radio)?
   (a) AT&T Wireless
   (b) Federal Dispatch Service
   (c) Nextel
   (d) Sprint Cellular (now 360 Communications)
   (e) None of the above

Satellite PCS Systems

27.1 Overview: The Role of Satellite Systems

The original concept that satellite systems had a role to play in personal communications (i.e., PCS) stemmed from several sources. Within the International Consultative Committee on Radio (CCIR—a standards body), satellite communications was seen as a means to offer direct service to users in less-developed countries lacking a strong telecommunications infrastructure. As with cellular and cordless services and products, technological progress facilitated size and cost reductions of satellite communications equipment by the mid-1990s, which led to the existing explosive development of new mobile services and equipment.

From 1995 to late 1997, a handful of companies spent billions of dollars honing technology, perfecting strategies, and closely watching each other in the race to develop and launch global, satellite-based mobile PCS systems. The technologies are different but the effects are the same: A businessperson from Chicago on vacation in the Himalayas should be able to pick up his mobile phone to call the home office, and a farmer in the Democratic Republic of the Congo in need of a replacement engine for a tractor should be able to call and order one from a parts manufacturer on the other side of the continent of Africa.

The companies involved in this development aren't creating a market for satellite-based mobile telephony as much as they are trying to answer a certain demand. Half of the world's population has no basic telephone service, and even in developing countries where there is money to pay for the service, it could take years to roll out the ground infrastructure to provide it. According to the International Telecommunications Union (ITU), 50 million people in 1998 who could afford a telephone (including government officials, merchants, manufacturers, traders, and farmers) can't get a connection. Some countries are attempting to solve the telephony shortage with wireless local loop, but with the satellite-based system, the result is immediate. A few companies, including Globalstar (which has made rural telephone service its focus), will offer fixed satellite phone service for use in rural areas with no telephone service. There, much as they would in a telephone booth, callers will pay only for minutes used. The question is this: how many people living in developing or third-world countries can afford to pay what may amount to a hefty portion of their salary for a telephone call? The solution to the need for telephone service in these countries is apparent but not always entirely feasible. One solution would be to set up a telephone booth which has satellite access, and to set it up as a service to an entire village. Perhaps then the village could afford the service.

27.2   Satellite Operating Modes

There are three major types of satellite operations available today: LEO (low Earth orbit) satellites, MEO (medium Earth orbit) satellites, and GEO (geostationary Earth orbit) satellites.

LEO satellites circle the Earth, usually in a polar orbit, at altitudes of 500 to 1000 mi. It takes around 48 to 56 LEO satellites to provide full coverage over the Earth. *Note:* Polar orbits provide better transmission coverage than equatorial orbits.

MEO satellites orbit the Earth at an altitude of around 6200 to 9400 mi. It takes around 10 MEO satellites to provide full coverage of the Earth.

GEO satellites orbit the Earth at an altitude of 22,400 mi. It takes around four GEO satellites to provide full Earth coverage. GEO systems are not usually used for pure voice systems because of the extensive amount of delay associated with GEO transmissions resulting from the high altitude of the satellites.

27.3   The Major Players and Market Estimates

The major players as of late 1997 were IRIDIUM, Globalstar, ICO Global Communications, Teledesic, and Odyssey Telecom International. These are the five early entrants into the developing field of satellite-based wireless phone service. Other companies are also in the process, as of early 1998, of applying for licenses to offer satellite-based communication services.

Globalstar has estimated the market for mobile and fixed voice services at 30 million potential subscribers. Odyssey put that number closer to 40 million subscribers; ICO at 50 million. These companies may be painting an optimistic picture of subscriber projections, and competition from small LEO satellites and GEO satellite systems that are providing mobile satellite communication to third-world countries is likely to reduce their potential subscribers. According to C.A. Ingley and Company, Washington,

D.C., a more realistic projection is probably closer to 14 million subscribers by 2004, growing to more than 25 million by 2009. These five companies are confident, however, that they will fill their allotted spectrum to capacity. Given the early demise of the IRIDIUM system (see Section 27.3.1.3), it is now very unlikely that the number of 25 million subscribers will even be reached.

All the major companies involved in satellite-based mobile telephony realize that the revenues they want to capture in the early years of service will likely come from international business travelers who would be willing to pay a premium to stay in touch anywhere in the world using voice, e-mail, fax, data, or paging services.

27.3.1   The IRIDIUM System

In this section, the IRIDIUM system is described in the past tense. That is because the first, most ambitious satellite PCS system to be developed and launched—IRIDIUM—met with an early demise in the year 2000, even though it was the first system of its kind to market. Other satellite PCS companies, IRIDIUM's would-be competitors, have learned much by watching the IRIDIUM system's demise just 2 short years after it was launched. It is still important for the reader to understand how the system was developed, because it was a very ambitious feat for its day, from both technologic and marketing perspectives. The downfall of IRIDIUM will be discussed in detail in Section 27.3.1.3.

While satellite personal communications had been proposed for some years, the development that brought it to widespread public prominence was Motorola's announcement in June 1990 to develop a worldwide digital satellite-based cellular personal communications system, to be known as IRIDIUM. Motorola's goal with this system was "to bring personal communications to every square inch of the Earth." The frequency band that Motorola used to offer this service was 1610 to 1626 MHz (1.6 GHz). In November of 1992, *Business Week* estimated that putting up the full IRIDIUM system would cost $3.4 billion.

Key: The IRIDIUM system was *not* intended to compete directly with conventional terrestrial wireless systems, but rather to *complement* them.

IRIDIUM would provide support as follows:

- To fixed users in regions currently without traditional, landline telecommunications infrastructure
- To mobile users in regions currently lacking mobile services
- To provide a backup service where mobile systems already exist

Speech, paging, fax, and data services were to be provided by the IRIDIUM system.

One proposed theory was that the IRIDIUM constellation of satellites would allow *solar-powered telephones* to be provided in remote villages, as well as allow international businesspeople to be confident of always being available, and able to contact high-level staff. Circa 1992 IRIDIUM was the most publicized system, and the one with the most technical detail made available in the public domain. Of the five major satellite system players, the IRIDIUM system was first to market: commercial service began on September 23, 1998.

27.3.1.1 IRIDIUM Operational Overview

The initial IRIDIUM proposal was based around a constellation of 77 small (about 320-kg) satellites, hence its name: The element IRIDIUM has an atomic number of 77. The IRIDIUM system satellites were to orbit the Earth in seven co-rotating planes, each plane containing 11 satellites that circle the Earth once every 100 minutes.

Key: In 1994, the implementation concept was revised to allow the number of satellites to be reduced to 66, but the name IRIDIUM stuck. The elements of the design described above remain, with one less co-rotating plane in place. Therefore, there were six co-rotating planes when the system was fully deployed, each containing 11 satellites, for a total of 66 satellites.

Within the IRIDIUM concept, satellites were to be accessed by portable, mobile, or transportable terminals using low-profile antennas.

The IRIDIUM system was to be more sophisticated than the systems of the competition because it employed intersatellite links, a technology never before used in commercial application. Satellites are internetworked to provide an orbiting backbone infrastructure capable of routing calls onward to other system users via intersatellite links or via gateway ground stations into the terrestrial PSTN. With this approach, a call was fed from an IRIDIUM handset to a LEO satellite 420 mi away, relayed from satellite to satellite, and then entered the PSTN through the ground station nearest the called party. While competitors claimed the technology was unproven, IRIDIUM executives disagreed, saying the military had used intersatellite links for many years in "black box" programs. This technology gave IRIDIUM the distinction of being the only system with the ability to operate if the ground portion of its system was wiped out by a disaster. In other words, a person in the United States could call a person in Japan using only the IRIDIUM network if they both owned IRIDIUM phones. Nearly 50 IRIDIUM satellites were in orbit as of late 1997, and the intersatellite link technology was tested and working. By late 1997, IRIDIUM had demonstrated the intersatellite link technology and the ability to launch satellites in multiple (north-south cross-links and east-west cross-links). IRIDIUM claimed that, because of the intersatellite link technology, it was the only system to offer true global coverage. See Figure 27-1 for a depiction of intersatellite links.

Figure 27-1
Intersatellite links used in the IRIDIUM system.

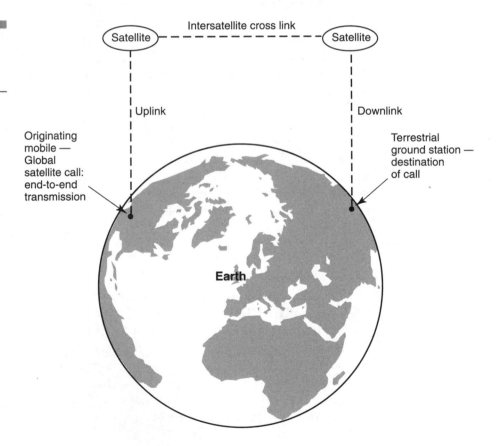

The full IRIDIUM constellation covered the Earth with a network of cells, each with a diameter of about 400 mi, enabling frequency reuse to be employed as in conventional wireless systems. A modified seven-cell reuse pattern was proposed. Each satellite would be capable of projecting a 37-hexagonal-cell pattern.

IRIDIUM pocket phones operated at 1.6 GHz with transmission power of 0.6 W. They used quadrature phase shift key (QPSK) modulation and a TDMA/FDMA access technique. They were based on GSM experience with similar features and protocols.

27.3.1.2 IRIDIUM and Interworking with Cellular Systems As mentioned previously, the IRIDIUM system would complement cellular, land-based systems. Therefore dual-mode phones were used so that *before* a call was connected via satellite, access to a local cellular network would first be attempted. This would avoid unnecessary burdening of the system with urban-originated calls, and would automatically offer the user least-cost routing. The presumption was that, obviously, a satellite-based call would be more expensive than a terrestrial, PSTN-based call.

27.3.1.3 The Demise of the IRIDIUM Satellite PCS System The individuals and organizations that invested so heavily in IRIDIUM, only to have the system begin to flounder by late 2000, must be frustrated not only with their financial losses, but also because such promising technology went unfulfilled.

In retrospect, it can probably be said that there were several things that caused the failure of IRIDIUM as a business.

First, IRIDIUM was beset by poor marketing on multiple fronts:

- The company placed heavy reliance on "international business travelers" as a chief source of revenue. Just how much revenue can be generated by international business travelers, in order to recover the investment of over $5 billion as rapidly as possible, in, say, 5 to 7 years?

- It didn't help that the system was not intended to compete with terrestrial wireless carriers, but rather to complement their service by providing coverage in the remote areas where a cellular or PCS signal could not be obtained (it should be noted that this is rapidly becoming a nonissue as of 2000). Even then, in a best-case scenario, the volume of traffic that would be generated by IRIDIUM users in very remote areas would be tiny in comparison to the volume of traffic generated by terrestrial wireless carriers, even in rural America!

- It was said that IRIDIUM also considered targeting very rural, remote second- and third-world areas to offer its service, sometimes in the form of "village phone booths." Again, this is another example of poor marketing. Most of the people living in these places couldn't afford to make an IRIDIUM phone call, not to mention another obvious fact: who on Earth (pun intended) would they call?

- The final instance of poor marketing came in the form of the IRIDIUM phones. Not only were they prohibitively expensive (around $3000), but they were large and bulky compared to today's portable wireless handsets offered by cellular and PCS carriers. To potential customers, it would almost be like taking a technological step backward. The oversized IRIDIUM phones with their bulky antennas resembled the "brick" phones used by cellular customers in the early to mid-1990s. The phones were also in short supply during the first 6 months of the company's launch, which hindered sales substantially.

Second, IRIDIUM's development and launch were unfortunately the victims of very poor timing. As the IRIDIUM company was completing development of its system and launching the last of its satellites in the late 1990s, the PCS carriers were rapidly building out their systems and aggressively marketing them as a substitute to cellular service. By late 1998, when IRIDIUM launched, the PCS carriers had gained firm competitive footing as a viable alternative to cellular service. Who would want to purchase "satellite" PCS and pay $5 to $7 per minute to place IRIDIUM calls, when the PCS carriers were effectively only charging around 10 to 12 cents per minute?

IRIDIUM had almost $500 million in annual interest expenses, a fairly high financial hurdle to cross. In May 1999, IRIDIUM hired Donaldson, Lufkin and Jenrette in order to restructure the company's debt. In July 1999, just 9 months after the IRIDIUM system went "live," it began to publicly discuss liquidation and bankruptcy. In August 2000, after frantically seeking "rescue" financing or even an outright buyer, IRIDIUM was jilted at the last minute by a prospective buyer, and gave up its efforts to keep the system operating. This cleared the way for Motorola to proceed with a "de-orbiting" procedure, in which all IRIDIUM's satellites would reenter the Earth's atmosphere and burn up.

IRIDIUM's would-be competitors, several of them still working on their own system development and market launches, have learned much from the IRIDIUM fiasco. Only time will tell which of them will reshape this market and redefine its potential.

27.3.2 Globalstar

The Globalstar concept was announced in October 1991 by a consortium including the following companies: Loral, Qualcomm, Aeorspatiale, Alenia, and Alcatel. The system is based on CDMA technology, pioneered for cellular and for satellite-based vehicle positioning applications by Qualcomm.

The Globalstar system will be composed of 52 LEO satellites, and the constellation will be connected to dozens of ground stations in countries around the world. Based in San Jose, California, Globalstar has forged partnerships throughout the world, giving countries and their wireless service providers an opportunity to invest in its system in exchange for spending a few million dollars to build a gateway in the country. Globalstar has also focused on the market in developing countries, realizing that traveling business executives alone cannot sustain a system in terms of cash flow and profits.

Globalstar's "bent pipe" system is the same technology currently used in geostationary satellites, but Globalstar will use the technology in low-Earth orbit to eliminate the voice delay. According to Globalstar executives, the beauty of the system is its proven simplicity: When a call goes from the ground to the nearest of Globalstar's 52 satellites, the satellite doesn't process the call like an IRIDIUM satellite, but merely relays it down to the gateway or ground station nearest the calling party, where it is processed and enters the PSTN. With this architecture (system on the ground), it's much easier to repair a problem and also incorporate changes when new technologies are developed. Globalstar plans to market the following services using its satellite constellation:

- One phone for cellular, PCS, or satellite calls
- Voice calling
- Short message service (SMS)
- Global roaming
- Facsimile
- Data transmission

Service availability for Globalstar is planned for late 2000, early 2001.

27.3.3 ICO Global Communications

ICO Global Communications, based in London, is a spin-off of a consortium that was known in 1992 as Inmarsat M. The commercial launch of the ICO satellite system in 1992 was the first step in the direction of commercial development of satellite-based wireless communication services. ICO has conceded its service wouldn't be ready for the early round. The company projects an early market entry in 1999 and a full rollout of services in the year 2000. The $4.6 billion system will employ 12 satellites in MEO orbit, 6430 mi above the Earth. Because the satellites are higher than LEO systems, but not high enough to suffer voice delay, fewer satellites are needed to "see" the Earth's surface. The path for an ICO call will be similar to that of a Globalstar call.

In May 2000, ICO was refinanced. This refinancing was led by Craig McCaw, a pioneer in the wireless industry and a major shareholder of Teledesic, another major satellite PCS player. The refinanced ICO was *merged* with Teledesic, and the new company is now called ICO-Teledesic.

27.3.4 Teledesic

The Teledesic satellite system represents the vision of wireless telecommunications pioneer Craig McCaw, the company's chairman. Teledesic's primary investors are McCaw, Bill Gates, Motorola, Saudi Prince Alwaleed Bin Talal, the Abu Dhabi Investment Company, and Boeing Corporation.

Teledesic is building a global, broad Internet-in-the-Sky™ network. Using advanced satellite technology, Teledesic and its partners are creating the world's first network to provide affordable, worldwide, "fiber-like" access to telecommunications services such as computer networking, broadband Internet access, interactive multimedia, and high-quality voice. The intention of Teledesic is to offer broad connectivity for businesses, schools, and individuals everywhere on the planet from day one. Teledesic is a private company based in Bellevue, WA, a suburb of Seattle.

Teledesic was founded in 1990, the same year that development of the IRIDIUM system was announced. In 1994, initial system design was completed, and an application was filed with the FCC to provide satellite PCS service. In 1997 the FCC granted the license to Teledesic, and the World Radio Conference designated necessary international radio spectrum for service as well. In 1998, Motorola joined the effort to build the Teledesic network. It is interesting to note that this is the same year that the IRIDIUM system was actually launched. In 1999, Teledesic completed a system agreement with Motorola, making Motorola a chief supplier to the Teledesic network. That same year Teledesic also signed a major launch contract with Lockheed Martin.

The Teledesic network is designed to support millions of simultaneous users. Using standard equipment, most users will have two-way connections that will provide up to 64 Mbps on the downlink and up to 2 Mbps on the uplink. 64 Mbps represents speeds more than 2000 times faster than today's standard analog modems! End-user rates will be set by service providers, but Teledesic expects rates to be comparable to those of future urban wireline options for broadband service. Teledesic will use 288 satellites, plus spares, in its system. It will operate in the high-frequency Ka band of the radio spectrum: 28.6 to 29.1 GHz uplink and 18.8 to 19.3 Ghz downlink. Service for the Teledesic system is slated to begin in 2004.

27.3.5   The Odyssey Telecom International System

Odyssey Telecom International, developed by TRW and Teleglobe, has said that both international (business) travelers and people in rural areas are important for their service. Odyssey would permit subscribers using only a pocket telephone to call any phone on Earth, from anywhere on Earth. Odyssey will also provide *fixed* wireless terminals for businesses and government organizations. The Redondo Beach, California–based company is proposing a 12-satellite MEO system in an equatorial orbit, at an altitude of several thousand miles. However, Odyssey suffered a major setback in the summer of 1997 when an unnamed leading investor withdrew from the program. According to an Odyssey spokesperson, the system is "still moving forward."

Odyssey is the wildcard in the space race. Although its competitors and some industry watchers are counting Odyssey out, the company said it is still moving ahead with plans for its $3.2 billion, 12-satellite MEO constellation. Odyssey's managing director, Bruce Gerding, insists that Odyssey is not in a "space race."

27.4   Implementation Issues for Satellite Systems

While all the involved companies are touting their services as "global," it will take a lot to get them there. Starting a global service means getting approval from more than 200 countries to sell the product, receiving permission to connect to the existing terrestrial network in those countries, and marketing and offering service to end users. Realizing they had more political clout as a group, the leading companies got together in 1996 to lobby the World Trade Organization and the ITU for the nondiscriminatory licensing of operators and terminals. In this context, *nondiscriminatory* means that all the companies could apply for the right to implement a terrestrial base station in a given country. The teamwork paid off, and it's now up to the companies to bargain with individual countries for licensing rights. One way they are securing licenses is to present themselves as a noncompetitive entity to the existing wireline and wireless carriers. All the companies are marketing a dual-mode handset that works with the customer's cellular network when a cellular signal is available. If a cellular signal is not available, the satellites come into play. Customers are charged by the satellite system operators only if a satellite is used to place the call.

The companies also face regional competition and perhaps roadblocks in some parts of the world, such as Asia, from smaller LEO and GEO operators who are trying to bring rural telephone service to their underserved populations.

A critical ITU meeting was held in March 1998 in Geneva, where representatives from countries and companies discussed the interconnect fees that will be charged on the satellite calls. The companies are pushing the FCC to ask the ITU countries to lower the fees, thus lowering the cost of the calls.

27.5   Customer Charges

While the per-minute prices for satellite calls have covered a very wide range, depending on what day of the month it is, in the end the charges for voice calls should all end up in the same ballpark—from $1 to $3 per minute, depending on where the call originates and terminates. The charges for packet data transmissions have apparently not been determined yet.

Mobile handsets in the satellite PCS marketplace will range in cost from $500 to $3000. The high-tech phones will be sold through local cellular providers and, in some cases, directly to large corporations willing to pay the price to keep their "road warriors" in touch.

27.6   Viability of Satellite PCS Systems

The handful of remaining satellite PCS companies have huge challenges ahead of them, in terms of proving they can launch and sustain successful businesses based on the premise of offering a product—satellite telecommunications services—that will be able to compete head-on with terrestrial wireless carriers. This issue has become self-evident with the untimely demise of the IRIDIUM system, which resulted in the loss of over $5 billion in investment.

Below is a list of the steps that the remaining satellite players will have to take if they want to establish themselves as valid players in the wireless communications industry of the twenty-first century:

1. Satellite PCS companies must focus their product offerings on more than voice services. In an era when data transmissions are now overtaking voice transmissions in the PSTN, any satellite PCS carrier would be making a grave mistake by focusing its product and marketing efforts solely on voice services. This is one of the reasons that the IRIDIUM system failed. Although all the satellite PCS carriers list their capabilities to provide "voice, fax, two-way paging, etc." in their tag lines and brochures, only one of these players has been truly creative in promoting itself via a marketing theme that's actually trademarked to protect its originality: Teledesic. The company's Internet-in-the-Sky slogan and outspoken focus on "broadband wireless" should carry it well in an era when the Internet is driving so much of the evolution of telecommunications today.

2. Satellite PCS carriers need to ensure that their phones (or "terminals") are comparative in look, feel, and *cost* to what can be obtained from a terrestrial wireless carrier. This is another area where IRIDIUM failed miserably. There is no incentive for anyone to purchase satellite wireless service if the phone (terminal) required costs 30 times what a land-based carrier would charge!!

3. Satellite PCS carriers need to show, in no uncertain terms, how their system adds value to what the consumer or business can purchase, when compared to terrestrial wireless carriers. To that end, it may also be time for the satellite players to decide whether it's really in their best interests to offer a service that does *not* compete directly with landline terrestrial services. Since the vast majority of potential customers will usually be in populated areas, maybe it's time for the satellite players to ensure that their customers will receive and use *only* a satellite signal when using the service. The approach taken by IRIDIUM (and at least one of the remaining satellite players) was to *not* compete with terrestrial wireless carriers, but rather to complement their coverage and system accessibility. If one of your competitors is receiving the majority of your customers' traffic before your own system is even considered a viable choice for call processing, that is both a design and business flaw in today's marketplace. This is obvious in an industry that is now hypercompetitive, where the service itself is well on its way to becoming a commodity.

The remaining satellite PCS companies must move very carefully, but also very aggressively, if they are to establish themselves in the wireless marketplace. By avoiding the mistakes of IRIDIUM and staking their own ground in the marketplace, they stand a chance to survive and thrive in the years ahead.

27.7   Satellite PCS Systems Summary

Although the companies hope for a quick turnaround on their investments, ultimately it will be the satellite PCS companies with the deepest pockets who will survive the battle for the mobile voice market. For the companies who have the backing to wait it out, the rewards will likely be great.

27.8   Test Questions

Multiple Choice

1. In 1992, the implementation concept for the IRIDIUM system was revised to allow for how many satellites?
   (a) 77
   (b) 55
   (c) 777
   (d) 66
   (e) None of the above

2. The IRIDIUM system was more sophisticated than the systems of the competition because it employed what type of technology?
   (a) Links to terrestrial Earth stations
   (b) Wideband CDMA
   (c) Intersatellite links
   (d) Interstellar links
   (e) None of the above

3. How many medium Earth orbit (MEO) satellites are needed to saturate the entire Earth with satellite coverage?
   (a) 100
   (b) 48–56
   (c) 10
   (d) 2
   (e) None of the above

True or False?

4. _____ The IRIDIUM satellite system was intended to compete with landline cellular systems.

5. _____ Low Earth orbit (LEO) satellite systems circle the Earth at an altitude of 500 to 1000 mi.

6. _____ Half of the world's population has no basic telephone service.

Broadband Wireless Technologies

28.1 Introduction to Broadband Wireless Technology

The demand for bandwidth is growing at an explosive rate as more businesses and individuals seek faster access to increasingly complex content on the Internet. Worldwide Internet use is expected to climb from 142 million subscribers in 1998 to more than 500 million subscribers by 2003, according to International Data Corp.

As recently as 1996, the only economical way for a business to receive and transmit high-speed data (defined here as data received at speeds of 1 Mbps and higher) was through the wired infrastructure built by the Regional Bell Operating Companies (RBOCs). For home data users, high-speed Internet access meant modems running at 28.8 or 56 kbps over telco twisted-pair copper cable.

With the limited availability of bandwidth from traditional telephone service providers and the worldwide deregulation of the telecommunications industry, a large market exists for broadband fixed wireless. Frost and Sullivan expects the broadband fixed wireless equipment market to grow at a compounded annual rate of 70% by 2004. Insight Research predicts broadband fixed wireless revenues to top $523 million in 2000; and the Strategis Group projects broadband fixed wireless revenues will climb from $11.2 million in 1999 to over $3.4 billion in 2003.

The "last mile" alternative to cable and digital subscriber line (DSL) in metropolitan areas, broadband fixed wireless is a viable technology of choice for suburban and rural markets, once these systems are fully deployed. Broadband fixed wireless systems are rapidly deployable, scalable, and have lower implementation costs than those of cable and DSL systems, which often require expensive infrastructure upgrades. Like DSL and cable, broadband fixed wireless is an always-on Internet access technology that can meet the growing demand for multimedia and voice applications.

28.1.1 Mobile versus Fixed Wireless

Until recently, the term "wireless communications" referred only to the use of mobile telephones. Consisting of highly cellularized architectures, telephone and PCS networks have one priority: optimizing the mobility of their users. No matter how much ingenuity is applied, these networks are *primarily* focused on providing digitized mobile voice connections.

While companies like AT&T, Microsoft, Qualcomm, and Motorola are working to overlay certain data capacities onto the traditional mobile wireless networks, these networks will still be configured for mobile or semi-mobile access using base stations with low-cost antennas. The cost per bit per second in traditional mobile wireless systems may be at a premium, and may reflect the mobility aspect of these systems.

Today, "wireless communications" refers to "mobile" as well as "fixed" technologies. With broadband fixed wireless, a highly directional, high-gain antenna is mounted at a fixed location. Because the site is stationary (hence the label "fixed"), the network's capacity can now be *redirected* to providing higher throughput. For mobile technology, the network's capacity would be needed to ensure *uniform coverage*. Broadband fixed wireless systems can use denser modulation schemes (i.e., CDMA) that require high signal-to-noise ratios. This enables data rates to easily reach 10 Mbps and higher, a 1000-fold increase from current mobile capacity. The result is that the cost per bit per second is *considerably* lower than that of mobile systems.

In this chapter, two maturing wireless technologies, which will challenge the traditional means used to deliver broadband services to residences and businesses are reviewed: Local Multipoint Distribution Service (LMDS) and Multichannel Multipoint Distribution Service (MMDS). Each of these technologies is being deployed with an initial focus on certain services and market segments, depending on the operator.

28.2   Local Multipoint Distribution Service (LMDS)

28.2.1   Overview

Local Multipoint Distribution Service (LMDS) is the broadband wireless technology used to deliver voice, data, Internet, and video services in the 25-GHz and higher radio spectrum (depending on licensing).

As a result of the propagation characteristics of signals in this frequency range, LMDS systems use a cellular-like network architecture, though services provided are *fixed*, not mobile. "Cellular-like" means that careful frequency reuse plays a key part in LMDS system design and operation.

The acronym LMDS is derived as follows:

Local denotes that propagation characteristics of signals in this frequency range (i.e., 28 GHz) limit the potential coverage area of a single cell site (base station). Ongoing field trials conducted in metropolitan areas place the range of an LMDS transmitter at 2 to 5 mi.

Multipoint indicates that signals are transmitted in a point-to-multipoint or broadcast method. However, the wireless return path, from the subscriber to the base station, is of course a point-to-point transmission.

Distribution refers to the distribution of signals, which may consist of simultaneous voice, data, Internet, and video traffic—in effect, multimedia transmissions.

Service implies the subscriber nature of the relationship between the operator (carrier) and the customer. The services offered through an LMDS network will be entirely dependent on the operator's choice of applications.

LMDS will occupy one of the broadest expanses of spectrum devoted to any one service, a total allocated bandwidth (in the United States) of 1.3 GHz surrounding the 28-GHz band, also known as the Ka band. That's *43 times* the amount of bandwidth obtained in December 1995 by the winning bidders in the first broadband PCS auctions. This bandwidth will be used to deliver broadband services in a point-to-point or point-to-multipoint configuration to residential and commercial customers. Baud rates will be in excess of 1 Gbps downstream (carrier to customer), 200 Mbps upstream (customer to carrier).

Proponents believe LMDS will be easily deployable, especially with no wires or cables to be laid or maintained. The economics of the business seem favorable. On the other hand, people are desperate for any broadband scheme to work. LMDS happens to be one of the most recent technologies to be considered a potentially viable broadband service option.

28.2.2 Spectrum Licensing

The FCC began the auction of 986 LMDS licenses on February 18, 1998, for operators who will offer LMDS in the 28-GHz and 31-GHz frequency bands; namely the A and B blocks. Each frequency block is allocated as follows:

A block (1150 Mhz of total spectrum)

- 28-GHz band: 27.5—28.35 GHz and 29.1—29.25 GHz
- 31-GHz band: 31.075—31.225 GHz

B block (150 MHz of total spectrum)
- 31-GHz band: 31—31.075 GHz and 31.225—31.300 GHz

These licenses can be used for wireless local loop (WLL), high-speed data transfer, video broadcasting, and two-way communications.

Incumbent local exchange carriers (ILECs) and cable television companies are restricted in their eligibility to own A block licenses in their authorized service areas. This restriction will stay in effect for at least 3 years from the date the license auctions were completed (March 1998). This restriction does not apply to B block license ownership. The issue at stake is that the ILEC would possibly block competition to a lower-cost local service provider by warehousing the LMDS license.

The LMDS auction concluded on March 25, 1998. Like the PCS auctions and market arrangements, LMDS operators will also operate within the structure of the basic trading area (BTA) system. The license fee for the top 25 urban BTAs ranged from $2.25 per POP (population) to $9.22 per POP for the A block; and from $0.27 to $2.01 per POP for the B block. The highest license fees topped out at $11.23/POP for Carlsbad, New Mexico, in the A block and $2.12/POP for Wichita, Kansas, in the B block.

28.2.3 Network Architecture and System Components

Various network architectures are possible within LMDS system design. The majority of system operators will be using point-to-multipoint wireless access designs although point-to-point systems and distribution systems can be provided within LMDS systems as well. Since LMDS will offer multimedia service, asynchronous transfer mode (ATM) and IP transport methodologies are practical options from LMDS transport infrastructures.

The LMDS network architecture consists primarily of four parts:

1. Network operations center (NOC)

2. Fiber-based infrastructure

3. Base station

4. Customer premise equipment (CPE)

The NOC contains the network management systems (NMS) equipment that manages large regions of the network. Of course, multiple NOCs can be interconnected for redundancy purposes.

The fiber-based infrastructure typically consists of synchronous optical network (SONET) optical carrier (OC3), OC12, and DS3 links; central office (CO) equipment; ATM and IP switching systems; and interconnections to the Internet and the public switched telephone network.

The base station is where the conversion from fibered infrastructure to wireless infrastructure occurs. Base station equipment includes the network interface for fiber termination, modulation and demodulation functions, and microwave transmission and reception equipment typically located atop a roof or a pole. A key functionality which may be present in different designs includes local switching. This function implies that billing, channel access management, registration, and authentication occur locally within the base station. If local switching is present, customers connected to the same base station can communicate with one another without entering the fiber-based infrastructure.

The alternative to this design simply provides connection to the fiber-based infrastructure. This forces all traffic to terminate in ATM switches or CO equipment somewhere in the fiber-based infrastructure. In this scenario, if two customers connected to the same base station wish to communicate with each other, they do so through a centralized location. Billing, authentication, registration, and traffic-management functions are performed centrally.

The customer-premise configurations vary widely from vendor to vendor. Primarily, all configurations will include outdoor mounted microwave equipment and indoor digital equipment providing modulation, demodulation, control, and customer-premise interface functionality. The CPE may attach to the network using TDMA, FDMA, or CDMA technologies. Choices for the customer-premise service interfaces will run the full range, including digital signal level 0 (DS0); plain old telephone service (POTS); 10BaseT Ethernet; unchannelized DS1; channelized DS1; frame relay; ATM(25); serial ATM over DS1, DS3, and OC3. The customer-premise locations will range from large enterprises (i.e., office buildings, hospitals, campuses), in which the microwave equipment is shared between many users, to mall locations and residences, in which single offices requiring 10BaseT Ethernet and/or two POTS lines will be connected. Obviously, different customer-premise locations will require different equipment configurations and different price points. See Figure 28-1 for a depiction of a typical LMDS system.

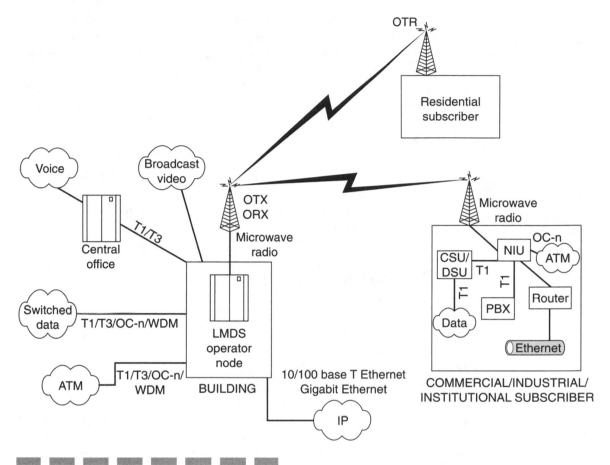

Figure 28-1 Typical LMDS network architecture.

28.2.4 Microwave Radio Operation in the 28-GHz Range

An area of continuing research for LMDS systems relates to microwave propagation behavior. The LMDS systems at 28 GHz are susceptible to rain effects, which can cause a reduction in the signal level and ultimately lower reliability of the system itself. The Consulting Committee on International Radiocommunications (CCIR) has rainfall attenuation estimation procedures; however, there is limited data and experience in small-cell, point-to-multipoint systems.

At the LMDS frequencies, multipath fading should not have a significant effect. First, LMDS frequencies are much more line-of-sight dependent, which means that shadowing and diffraction (signal bending) do not occur as often. Second, LMDS systems have customer antennas located high on rooftops, whereas cellular and PCS systems typically have customer antennas within 6 ft of the ground. The height of the customer (premises) antenna plays a large role in reducing multipath effects. Third, the LMDS antennas are highly directional (pointing to a single cell site), where cellular and PCS base station antennas have either omnidirectional or loosely sectorized characteristics. Using directional antennas reduces multipath effects. Fourth, in cellular and PCS systems the customer (antenna) may be moving, whereas LMDS antennas are fixed on a rooftop. Once an antenna becomes fixed, installers can choose better locations on the rooftop, which leads to improved performance.

28.2.5   LMDS Applications

There is no consensus regarding what LMDS will ultimately be used for since it can offer such a variety of applications. Some in the telecommunications industry say that the market for LMDS will vary tremendously between BTAs.

Industry observers see three major opportunities for LMDS deployment. One option is for LMDS to play a large part in the interactive TV market. One firm, CellularVision, is already using LMDS in the 28-GHz band to deliver the equivalent of 49 channels of cable TV service to parts of New York City that have resisted deployment of coaxial cable. CellularVision also has plans to gradually introduce two-way services such as video-on-demand and high-speed Internet access starting at 550 kbps.

The second potential market is high-speed data. LMDS cannot only provide access to the Internet, it can also be used to interconnect local area networks (LANs), create campus area networks, and serve as a medium for SONET. These capabilities enable LMDS operators to focus on business customers *first*—those best able to afford 28-GHz subscriber equipment before the development of a mass market.

The third potential application for LMDS is basic voice service. Because it can handle thousands of voice channels, LMDS' potential for providing voice service should not be discounted. It is highly unlikely that LMDS systems would be built out just for POTS, but voice could represent a good source of incremental revenue.

Because mobile access is problematic at the 28-GHz frequency band, LMDS will not compete with cellular and PCS providers.

28.2.6 LMDS Standards

As LMDS wireless access systems evolve, standards will become increasingly important. Standards activities currently under way include activities by the ATM Forum, the Digital Audio Video Council (DAVIC), the European Telecommunications Standards Institute (ETSI), and the International Telecommunications Union (ITU). The majority of these methods use ATM cells as the primary transport mechanism.

28.2.7 Deployment Strategy and Benefits of LMDS

Point-to-point *fixed* wireless networks have been commonly deployed to offer high-speed dedicated links between high-density nodes in a network, for example, a microwave radio-based cellular or PCS fixed network. More recent advances in point-to-multipoint technology offer service providers a method of providing high-capacity local access that is less capital intensive than a wireline solution, faster to deploy than wireline, and able to offer a combination of applications.

Additionally, because a large part of an LMDS wireless network's *cost is not incurred until the customer premises equipment* (CPE) *is installed,* the LMDS network service operator can time capital expenditures to coincide with the signing of new customers. This represents a radical departure from the way that the incumbent telcos have operated over the last 100 years, where miles of cable plant and CO-based equipment had to be purchased and installed prior to providing service to customers. LMDS provides an effective last-mile solution for incumbent service providers, *and* can be used by competitive service providers (i.e., CLECs) to deliver services *directly* to end users—customers.

The benefits of LMDS can be summarized as follows:

- Lower market entry and deployment costs.
- Ease and speed of deployment. Systems can be deployed rapidly with minimal disruption to the community and the environment.
- Rapid realization of revenue as a result of rapid system deployment.
- Demand-based buildout. LMDS offers a scalable architecture employing open industry standards, which ensures services and coverage areas can be easily expanded as customer demand warrants.

■ Cost shift from fixed to variable components. With traditional wireline systems, most of the capital investment is in the infrastructure, but with LMDS a greater percentage of the investment is shifted to CPE, which means an operator spends dollars only when a revenue-paying customer signs on.

■ There is no "stranded capital" when customers churn off the system.

■ Cost-effective network maintenance, management, and operating costs.

28.2.8   LMDS Summary

To date, three major carriers have deployed LMDS systems in one form or another: Nextlink, Teligent, and Winstar. Nextlink owns 95% of the LMDS spectrum in the top-30 markets nationwide. Vendors have not finalized equipment specification or pricing, but early indications show a preference for ATM-based solutions similar to those employed by Teligent and Winstar. As with all solutions in the LMDS frequency range, a high degree of cellularization is required for operators building out LMDS systems. Cell size is about 2 mi in radius, or 12.6 square mi of total coverage. Experts estimate higher-frequency licenses, such as LMDS, require 5 to 10 times more investment for initial systems than MMDS.

28.3   Multichannel Multipoint Distribution Service (MMDS)

28.3.1   Overview

Multichannel Multipoint Distribution Service (MMDS) is often referred to as "wireless cable." It is a unique wireless technology for delivering high-speed Internet access. Its development and impending use stem from the Telecom Act of 1996, which sought to increase choice and competition in all aspects of telecommunications.

Ultimately, the deployment of nationwide MMDS systems will provide a choice to many Americans whose options for high-speed Internet access are currently limited to DSL controlled by their local telephone company or a "data" CLEC, or cable dominated by AT&T or Time Warner. The roster of potential subscribers to MMDS service includes:

- Residential customers, especially those in rural areas where it is cost-prohibitive for cable companies to build systems, or for telephone companies to supply DSL service.
- All schools, from kindergarten through higher education, need data.
- Telecommuters. Individuals in this growing sector need high-speed connections to communicate with their offices.
- Branch offices. Small to medium-sized businesses must have a network to transmit data between their offices.
- Businesses in "urban sprawl" areas. Businesses that are located away from the primary business district, in areas that are years away from having either DSL or fiber nearby, could access high-speed data via MMDS technology.

28.3.2 MMDS Spectrum and Its Development

During the 1960s, 1970s, and 1980s, the FCC allocated approximately 200 MHz of radio spectrum at the 2.1-GHz and 2.5- to 2.7-GHz frequencies for television transmission. The channels allocated to MMDS have traditionally been used to provide multichannel video programming service that is similar to cable television.

Originally, the MMDS wireless spectrum consisted of 33 analog video channels, each with a bandwidth (channel spacing) of 6 MHz. With new digital technology and the new competitive mandate (i.e., Telecom Act of 1996), the FCC greatly increased the flexibility of this frequency band beyond simple video, to full two-way digital communications, excluding only mobile communication services. In two separate rulings in 1995 and 1998, the FCC allowed for digital transmission utilizing CDMA, quadrature phase shift keying (QPSK), and quadrature amplitude modulation (QAM) schemes. This evolution can yield up to 1-Gbps total raw capacity for this frequency band, for send and return transmission, from multiple sites within a 35-mi-radius "protected" service area.

The evolution of video technology from analog to digital enabled newly licensed MMDS operators (circa 1998) to convert the 33 analog

MMDS channels into 99 digital data streams, each transmitting at 10 Mbps. As a result, MMDS operators can deliver up to 1 Gbps of capacity from a *single transmitter* to provide broadband access to data, voice, and Internet services since digitization of this band was allowed by the FCC.

MMDS shares the 2500- to 2690-MHz band with licensees such as colleges and universities, local school systems, and religious educational institutions, all of which use the spectrum to broadcast educational programming and lease their excess spectrum capacity to companies such as Sprint's Broadband Wireless Group (BWG).

In 1997, the FCC allocated a large portion of MMDS frequencies for the use of bi-directional high-speed data transmission. Subsequently, two major telecommunications companies, Sprint and MCI WorldCom, made major investments to gain the majority control of the MMDS spectrum in the United States.

The spectrum is utilized in omnidirectional fashion from a central antenna and may be cellularized as necessary. A major advantage in using this frequency band is that the physical need to cellularize is much less than in other major frequency bands. With appropriate terrain characteristics, a single MMDS cell (base station) can cover a 35-mi radius, or 3850 square mi!

28.3.3   Signal Propagation

The single most important factor in the structure and economics of the infrastructure required for a given wireless frequency band is the band's propagation characteristics, in other words, how far the signal reach is at a given power level under given terrain, foliage, and weather conditions. For example, as was pointed out earlier, the 800-MHz frequency band is ideal for cellular mobile wireless propagation because of a combination of factors: it is easily absorbed, it tends to be line of sight, it is easily reflected, and it has a short signal wavelength. In the MMDS frequency band, line of sight is a dominant requirement—the path between transmitter and receiver must be substantially unobstructed. If the terrain is hilly or if foliage is dense, line-of-sight opportunities must be multiplied through cellularization. Beyond a certain distance, a receive antenna of a given size and gain is unable to receive the signal. Larger and larger receive antennas could be put into place, but this obviously would severely limit the willingness of homeowners and businesses to use the service.

Key: Generally speaking, an antenna size of 18 in or less is essential for a service to gain wide acceptability.

At certain frequency bands, attenuation can be greatly exacerbated by effects of weather. At higher frequency bands, such as 24, 28, and 38 GHz, we have seen that the wavelengths are short enough that raindrops can actually present line-of-sight obstacles and greatly attenuate the signal. This is not an issue with the MMDS frequency band. Because of its longer wavelength, rain represents no obstacle to a properly engineered system. In its prior incarnation transmitting TV signals, there are MMDS systems with *years* of continuous transmission history that have not been affected by rain.

Like broadcast television, MMDS is transmitted from towers, usually located on a mountain or tall building, to special antennas affixed to residences or businesses throughout a market. Ideally, service is delivered using land-based radio transmitters positioned at the tallest feasible location in a metropolitan area. For example, Sprint Broadband Wireless Group has installed its transmitter in Chicago on top of the 1454-ft-tall Sears Tower, one of the top-five tallest buildings in the world. In Phoenix, Sprint's transmitter is on top of South Mountain.

28.3.4   System Components

28.3.4.1   CPE   At the customer's premise, the following equipment is used to provide for MMDS service: a customer's PC with an Ethernet card, or a customer's LAN switch, is attached to a wireless broadband router (WBR). This device converts the data from the PC or LAN to a signal suitable for transmission over RF facilities. In addition, it receives and sends intermediate frequencies (IFs) to the next device.

Via coax cabling, the WBR modem is connected to a transverter, also referred to as a transceiver, which converts the IF to radio frequencies and passes them to the antenna. The transverter and antenna are generally one piece of equipment, and are typically mounted on a customer's roof or on a mast to gain line-of-sight visibility to the operator's transmitter site.

28.3.4.2   MMDS Base Station   The base station is usually located at a very high geographic location. Transmit antennas are omnidirectional when the base stations are in central locations. When frequencies need to be reused, the base station (antennas) can be sectorized. In this scenario, multiple receive antennas are deployed to account for space diversity. This allows for better reception of the weaker signals coming from the CPE RF equipment. A waveguide is used to transport the transmit and receive signals to a downconverter for the receive side and a transmitter for the transmit side. Low-noise amplifiers are sometimes used to boost the receive signal before it gets to the downconverter. A gateway router is used to provide the connection to the Internet and other content servers. Finally, a fast-Ethernet switch is used to interconnect the various components. In the near future, this component will probably be upgraded to a gigabit Ethernet connection, if this migration hasn't occurred already.

28.3.5   How MMDS Works

Following is a step-by-step illustration of how MMDS works:

- Your computer sends a request for data or a web page to the MMDS modem.
- The MMDS modem sends the data request to the receiver/transmitter on the user's rooftop.
- The receiver/transmitter sends the data in a 2.1-GHz signal at speeds up to 10 Mbps to the MMDS operator's receive/transmit tower.
- The tower relays the data request through the network to the Internet service provider (ISP) facility.
- The ISP receives the request and retrieves data either from its servers or from the Internet over its own high-speed backbone connection.
- The ISP then returns the data via the network to the receive/transmit tower. The MMDS operator's transmit site sends the data in a 2.5-GHz signal at speeds up to 10 Mbps to the receiver on the user's rooftop.
- The roof-mounted receiver relays information to the MMDS modem. The modem passes the information to a stand-alone PC, LAN switch, or multiple users in a LAN all in seconds.

28.3.6   MMDS Benefits and Advantages

MMDS technology offers many advantages for carriers who deploy these systems, and for potential residential and business customers as well.

- *Spectrum.* MMDS represents a new generation of an already established and proven wireless service. For over 30 years, the radio-wave wireless spectrum has been used to transmit television signals. Now that MMDS' traditional spectrum has been digitized, it has been adopted to beam data and Internet traffic.

- *Interference.* Using the *licensed* spectrum, MMDS operators can deploy high-speed wireless LANs on secure channels without interference concerns associated with using unlicensed spectrum from other users or other wireless signals. Plus, unlike other wireless technologies or satellite transmissions, the signal performance of MMDS is very resistant to rain fade.

- *Coverage.* A single MMDS cell base station can serve a 35-mi-radius coverage area (over 3800 square mi). Higher-frequency wireless technologies such as LMDS, in contrast, cover an area with only a 2- to 5-mi (maximum) radius. It would therefore take over 136 LMDS cell base stations to cover the same areas as just *one* MMDS cell base station.

- *Infrastructure.* Radio waves ride the air, a very inexpensive medium to use as a system interface. The frequencies are what cost money. According to Motorola, it costs around $25,000 per mile to lay new two-way hybrid-fiber coaxial cable. In New York City, the cost to lay fiber-optic cable can run as high as $3 million per mile! Deploying a wireless solution for broadband transmission is much more cost effective than using cabled transmission systems. The cost of installing a single MMDS tower is projected to be around $6 million. This cost is equivalent to around $2000 per square mile, significantly less than the cost of *wiring* a square mile. Most importantly, installation is easy, "immediate," and not dependent on the quality and availability of phone lines.

- *Availability.* MMDS technology that's been licensed to major providers (e.g., Sprint) will be available soon, once the operators complete buildouts of their systems. MMDS has the potential to reach 30 million U.S. homes. Conversely, it will take years and billions of dollars for the RBOCs and cable multiple system operators (MSOs) to retrofit thousands of miles of outdated, poor-quality network cable plant, which requires removing line concentrators and loop conditioning.

- *Scalability.* To meet future growth of the customer base and bandwidth requirements, MMDS operators can expand the already vast capacity of MMDS systems using two methods:

1. By sectorizing transmitter stations/antennas
2. By cellularization or reusing the same frequency with multiple transmission sites

28.3.7   MMDS Summary

Once MMDS is available to the public as a realistic alternative to wire-based high-bandwidth technologies, mainly used for Internet access, a whole new dimension to the competitive landscape will emerge. MMDS is already deployed in nearly 50 markets across the United States. It is ideally suited for areas lacking access to DSL and cable.

MMDS will deliver speeds up to 100 times faster than traditional dial-up connections. To put that speed into perspective, consider that a 10-Mbps MMDS wireless modem could download the $3\frac{1}{4}$-hour movie *Titanic* in 7 minutes, 23 seconds. A DSL or cable modem would take 9 minutes, 14 seconds to download the movie. A DS1 connection would take 49 minutes, 20 seconds. ISDN would need 9 *hours*, 14 minutes. And finally, a 28.8-kbps modem would need a whopping 42 *hours*, 30 minutes, or 13 times longer than the movie itself.

If MMDS service is launched and proven to be a *reliable* mode of Internet access, it will present a credible, unprecedented challenge to the "wired" status quo for Internet access in the telecommunications world of the twenty-first century.

28.4   Test Questions

True or False?

1. _____ The key catalyst for evolution of the MMDS industry revolved around the conversion of video technology from analog to digital. This enabled newly licensed MMDS operators to convert traditional TV channels to digital data streams, offering data rates at 10 Mbps.

2. _____ LMDS will occupy one of the broadest expanses of radio spectrum devoted to any one service.

3. _____ There is no consensus regarding what LMDS will ultimately be used for since it can offer such a variety of applications.

4. _____ MMDS base stations can cover an area equivalent to 2 square mi.

Multiple Choice

5. The LMDS network architecture consists primarily of four parts. Which item below is *not* one of those four components:
   (a) Network operations center (NOC)
   (b) Satellite transponder
   (c) Base station
   (d) Fiber-based infrastructure
   (e) CPE

6. What technology is sometimes referred to as "wireless cable"?
   (a) LMDS
   (b) ATM
   (c) MMDS
   (d) Satellite PCS
   (e) CDMA

The Future of
Wireless Telephony

29.1 Intense Competition in the Wireless Marketplace

As PCS systems come on line, each geographic area of the United States may have up to seven or eight wireless carriers offering service to the public by 2002! That's a three- to fourfold increase from the traditional two cellular carriers who have offered service from 1983 to 1996. By the end of 2000, it is possible that all of the following wireless providers will be offering service to the public:

■ Two cellular carriers (A and B band)

■ Two MTA broadband service providers (A and B band)

■ Up to four BTA service providers

Note: The list above does *not* include the narrowband PCS licensees who will also be ultimately operating both regionally and nationwide, offering two-way data messaging and paging services.

According to standard economic theory, when there are many service providers vying for customers within one market, the prices for the service will trend downward, and the service should become more reliable and feature-rich. Wireless service is no different. In this case, the consumer should win in the end.

29.2 Multiple Options for Interconnection to the PSTN

The Telecommunications Act of 1996 is supposed to open all telecommunications markets to competition. This includes local exchange service, intra-LATA toll service, inter-LATA/interstate long distance service, and cable television service. Landline carriers of all types will eventually be allowed into each other's markets. Just as the *wireless* world will have intense competition over time because of the entrance of PCS companies into the market, the same onset of competition will eventually apply to the landline carriers. Because of this impending competition, service offerings by landline companies should become cheaper and more expansive. This will ideally benefit wireless carriers in the form of multiple, less expensive options for interconnection to the PSTN. The only caveat for wireless carriers when choosing among a number of companies offering local and long distance interconnection is to ensure that the companies offering interconnection services provide *reliable, quality service.* It will behoove wireless carriers to *not* sign any long-term contracts for interconnection services for these reasons:

■ If it turns out that a certain carrier is not offering quality service, a wireless carrier would not want to be tied down to a long-term contract or else would want a solid escape clause in the contracts.

■ Because there will likely be multiple carriers offering interconnection, a wireless carrier may want to retain the option to keep trying to interconnect with *various* PSTN companies *until* they find a company that offers quality, low-cost interconnection services.

29.3   Wireless PBX

29.3.1   Overview

The wireless PBX is also known as an *in-building cellular coverage system.* The service is offered to businesses by cellular carriers, and ultimately PCS carriers as well. A wireless PBX supplements existing telephone systems (i.e., PBXs) by utilizing wireless handsets that communicate internally with existing PBX lines. Employees are provided handsets that work in conjunction with the phones on their desks. For example, when the phone on someone's desk rings, the wireless PBX handset that is in that person's pocket (or sitting in a cradle on the desk) also rings. The employee then has the option of speaking on the regular PBX deskset or communicating through the wireless PBX handset. The idea is to give the employee the flexibility to roam throughout the building with only one telephone line. The handsets are also *feature-rich* because the cellular handsets are capable of using all the calling features that a business uses through the regular PBX.

Examples of business environments that would be ideal for the use of a wireless PBX system are hospitals (so that doctors can travel through the hospital and not miss any calls) and warehouses. Any business that requires its employees to work throughout the confines of a large building is a perfect candidate for a wireless PBX system. Wireless PBX systems are usually custom-designed, and the capacity of these systems can range in size from four handsets up to around 300 handsets. The cellular handset of a wireless PBX can also act as a regular cellular phone on the external cellular system.

If employees leave the designated in-building coverage area and go out-side, they can use the external "macro" cellular system.

If a call is initiated on the internal wireless PBX system and the caller travels outside the building (outside the wireless PBX designated cover-age area), the call will be dropped and would need to be reinstated man-ually on the external cellular system. Calls-in-progress are not handed off from the in-building system to the external "macro" cellular system.

29.3.2 Frequencies

With respect to the cellular (or PCS) frequency that's used by a wireless PBX, the in-building system provider must obtain authorization from the local cellular carrier selling the service to use a portion of its radio spectrum. Once the cellular carrier has granted the company the right to use a portion of its spectrum, the system can be implemented.

29.3.3 Wireless PBX Operation

A control unit, also known as a *controller*, is directly connected to the business's PBX, or it can be connected to centrex lines as well. The con-trol unit communicates to base stations (cells) that are placed throughout the building. The base stations handle all call handoffs for the wireless handsets, within the building.

Along with the control unit, the wireless PBX requires a scanning sta-tion, or "sniffer." The sniffer scans the external macro cellular system to determine which radio frequencies are being used, seeking the least used frequencies. The sniffer then communicates to the controller which chan-nels are being used by the outside system so these channels are not made available in the channel pool of the internal system (wireless PBX). The purpose is to keep the wireless PBX from interfering with the external cel-lular system (by cochannel and adjacent channel interference). The sniffer performs its task on the basis of an update window which can be set to add or delete available channels anywhere from every hour to every 48 h.

29.4 Wrist Phones

With the launch of PCS wireless systems, many industry sources predict-ed that the day would not be far off, say within 10 years, when wireless carriers would begin marketing wrist phones. The prediction that PCS handsets are supposed to be smaller and lighter than cellular handsets, coupled with the fact that Motorola is continually marketing smaller and smaller wireless phones, indicated that the wrist phone may indeed be a reality in the near future.

In March 1998, the company that produces the Swatch line of products unveiled a wrist phone. It is only a matter of time before the bigger players (e.g., Lucent, Motorola) develop and introduce their own versions of the wrist phone. Such phones would almost have to be voice-activated. Apparently, the public isn't ready for this leap yet; as of 2000 these phone watches have never really caught on.

29.5   Landline Replacement

One of the eventual goals of most cellular and wireless companies is to completely displace the need for a landline telephone network. Some industry experts even theorize that the PSTN may eventually be used solely for bulk data transmission systems.

One of the goals of PCS companies is to provide high-quality wireless service at such low rates that people will be tempted to simply do away with their traditional residential phone service, usually provided by a Baby Bell, and switch to wireless service completely. The key element to making this happen is for the PCS companies to make absolutely sure that their rates are competitive with local landline service. In other words, rates for PCS service must be as low as or lower than those charged by the landline companies. When this time comes, as it surely will, there will be a marketing war as each side jockeys to win the competition for a huge, lucrative market. Traditional telephone companies will not sit idly by while a huge, traditional source of revenue (that they've always taken for granted) is threatened. The consumer will be the winner in this competition, as each side will ultimately offer expanded services and lower prices.

29.6   The Development of a 3G Wireless Standard

There is a move afoot by global standards bodies to come to a decision on a standardized platform—a single wireless technology that will be the springboard for the development of advanced wireless services into the twenty-first century. The move describes efforts to develop a third-generation (3G) digital wireless technology that is evolved from the major existing digital wireless radio technologies: TDMA, CDMA, and GSM.

The first-generation wireless service was analog services. In the United States, this refers to the development and deployment of the AMPS analog cellular standard. The second-generation wireless service describes the all-digital PCS services—for example, TDMA, CDMA, and GSM.

Therefore, the third-generation wireless technology will be an enhanced version of one of the existing three digital wireless technologies, perhaps even a combination of one or more of these technologies. The 3G service will offer advanced services and features compared to those that are available circa 2000. One example of this development is the so-called W-CDMA technology, a design which employs a 5-MHz carrier compared to the current 1.25-MHz carrier. The international standards bodies are grappling with the issue of deciding which digital technology should serve as the springboard for *the* next-generation wireless service.

29.7   Billing Zones

Locator technology that is currently being developed will allow operators to designate very specific billing zones, even small areas such as user's homes. The FCC E-911 mandate is pushing location-sensitive billing into the minds of carriers and vendors. Location-based billing might increase revenues by lowering bills, which would make carriers more competitive. Wireless carriers could set rates that are as low as land-line costs for each subscriber's home. If this is done, wireless companies can compete with the local exchange carriers, says John Simms, CEO of SCC, a company that offers wireless E-911 services. With the excess capacity that PCS offers, location-sensitive billing supplies operators with a tool that encourages use of the mobile phone. For example, if an employee of a company were using a company-provided wireless phone, he could be charged one rate when making calls from his home and a (cheaper) bulk rate for calls made in or near his office. Carriers could optimize revenues by encouraging traffic where they have a surplus of capacity. For example, wireless operators could lower rates in low-usage areas and at low-usage times, such as 3:00 P.M. in the suburbs. They could also increase rates in high-usage areas at specific times, such as certain highways during rush hour. Market segmentation could also be a driver for location-based billing. For example, carriers could analyze usage patterns and determine the kind of pricing they need. Reduced rates within sport stadiums could encourage wireless use instead of pay phone use (again, this demonstrates an opportunity for wireless carriers to compete with the LECs).

Some carriers can offer some location-sensitive billing services today without using locator technology. For example, in an area with a heavy concentration of cell sites, such as a neighborhood with many wireless subscribers, operators can define a cell or set of cells where users receive a discounted rate.

29.7.1 Convenience and Service Intelligence

Locator technology might even help to make cell phones something like a wireless personal assistant. They could be programmed, for instance, to recognize the neighborhood when you drive near your laundry, and beep to remind you to drop in to pick up your shirts. Or, suppose you run out of gas while driving in a strange neighborhood, or lose a filling while chewing on a restaurant's tough steak. If you could punch in the word GAS on our cell phone and automatically ring up the nearest gas station, or dial the letters DOC to locate a nearby dentist to replace your filling, that might be a welcome service. At this time, this type of technology is still on the proverbial drawing board, but it is coming, and the wireless industry is hard at work developing it to full potential.

29.7.2 Mobile Phones as Tracking Devices

Law enforcement sees phone locator technology as a way to help them track criminals. But that's a prospect that gives some civil libertarians the chills. According to Andrew Blau, director of communications policy for the Benton Foundation, it is a classic societal trade-off: "On the one hand, people who dial 911 want help. They want to be found. But when locator technology makes your cell phone a de facto tracking device, that's a level of potential surveillance most Americans will find extraordinary and troubling." A proposal pending in Congress in 2000 forbade wireless carriers from making any use of locator technology apart from 911 calls, without the customers' expressed consent.

29.8   Wireless Local Loop

In a (public) telephone network, wireless local loop (WLL) is a generic term for an access system that uses a wireless link to connect subscribers to their local exchange in place of conventional copper cable. Using a wireless link shortens the construction period and also reduces installation and operating costs. Peter Nightswander of "The Strategis Group" defines WLL in the United States as "sort of the same as overseas: any type of wireless link to a subscriber where you use the system inside the home and/or outside the home. It could be fixed or mobile." In the WLL context, "fixed" means that WLL service is only functional inside the home, and maybe outside the home within a few hundred meters at the most.

29.8.1   The Market for WLL Systems

The wireline telephone service penetration rate differs greatly from one country to another, with some areas of the world *not even* having one telephone per 100 population. There is an urgent need to eliminate the backlog for telephone service—now estimated to be over 50 million lines—in order to make global communications a true reality for all people. As countries move ahead with plans to privatize telecommunications and introduce competition, investments in telephone network facilities are continually increasing. Subscriber access lines account for a relatively large share of the total investment in outside plant and equipment. If the cost of the local loop could be reduced, it is believed that the telephone service penetration rate would advance dramatically with one stroke. The last 300-m segment to each subscriber's premises accounts for a large share of the total cost of installing subscriber lines. Using a wireless link for subscriber access makes it possible to initiate service in a very short period of time, because the installation of the wireless facilities involves only a limited amount of small-scale construction. Wireless links are especially effective for the last 300 m. Also, as a result of the explosive spread of wireless service in recent years, wireless equipment prices have dropped dramatically. The cost of installing a WLL system for subscriber access lines is now virtually the same as that of copper cable, though it can vary depending on local conditions.

29.8.2 Implementation Requirements

The following list spells out the requirements and considerations that must be adhered to when considering implementation of a WLL system:

1. *Quality.* Since a WLL system serves as the access line for fixed telephone sets, it must provide the same level of quality as conventional telephone systems with respect to such issues as speech quality, grade of service (GOS), postdial delay, and speech delay. In addition, since radio waves are used, careful consideration must be given to privacy—protection of confidentiality and terminal authentication.

2. *Short construction period.* In order to serve as a viable alternative to wireline telephone service, the construction time frame required to install WLL systems must be considerably shorter than the time required to terminate copper cable to subscriber's premises.

3. *Low cost.* The overall cost to install a WLL system, especially in developing nations, must be low. This includes equipment, construction, and maintenance costs.

4. *Absence of interference.* A WLL system must not cause any interference with the operation of other existing wireless systems, such as microwave communication and broadcasting systems.

5. *High traffic volume.* A WLL system must be able to support a larger traffic volume per subscriber than regular mobile communication systems.

29.8.3 Marketing and Deployment Considerations

Wireless local loop may represent the possibility of one day truly competing in the local telephone market and taking competition to new heights. But as WLL is still in its infancy, WLL in the United States is anything but a sure win. In fact, if the LMDS (Local Multipoint Distribution Service—spectrum that is expected to play a role in future WLL applications) auction is any indication of what is to come, the WLL "underdogs" still have some skeptics to win over. In the United States, the marketplace and players in the WLL industry are still undecided. Most carriers and vendors have not yet even begun serious development efforts with regard to WLL systems. One reason for the slow activity simply may be that U.S. carriers are trying to understand exactly what WLL means domestically. Most of today's predictions for opportunity refer to the international market. Traditionally, WLL has been used to describe international efforts at offering wireless service as a basic telephony alternative in developing countries, countries that have little or no telecommunications infrastructure in place. The marketing waters for U.S. deployment are murky partially because the United States has the largest telecom infrastructure in the world, with more than 100 million access lines in place. That means there's not much need for wireless companies to build out basic telephone service, except in a few remote rural areas. However, industry optimists tout WLL as the replacement for landline service in the residential market. But there are many obstacles that WLL must overcome in order to realistically be considered as a viable alternative to landline-based telephone service. Although people don't necessarily like their local phone companies, at least they know whom they are dealing with and know they're getting reliable phone service that works. *Wireless carriers can't yet offer that 100% reliability!* Also, only 1.4% of the total minutes of telecom talk time in the United States is transmitted over a wireless device, which means the wireless industry has a lot of work to do in terms of closing that gap. The bottom line is that wireless carriers need to offer highly reliable, uninterrupted phone calls with high levels of quality, and right now the cellular and PCS industries are light years away from that. There is one opportunity that may present itself as a means for WLL to enter the public marketplace as technology advances: as a second or third phone line in the home. As demand for additional phone lines for fax and Internet applications increases, eventually WLL may offer a competitive solution for these applications in the residential market. Cost is another concern. At a WLL seminar in early 1998, Craig Farrill, a vice-president at Airtouch Communications, asserted that the current infrastructure costs for WLL are too high to justify a solid business case to build out WLL systems at this time. Farrill contends that the costs need to be cut in half for WLL to really take off as an industry. Some experts also believe that there are regulatory hurdles to overcome, that today's policies do not encourage wireless carriers to get into the local loop. Despite regulatory obstacles, a few vendors such as Lucent, Nortel, and Tadiran are already touting their WLL and fixed wireless solutions in the United States through trade shows and press releases. But for many vendors, the investment in the U.S. market is still too risky and uncertain, because there is always the risk that the market won't deliver.

29.9　Current Assessment and Industry Predictions

29.9.1　Wireless Industry Growth Projections: 1999 and Beyond

As of late 1998, dozens of PCS carriers have risen above a multitude of problems to gain a solid footing in the competitive marketplace, and they are giving the incumbent cellular providers strong competition. Problems such as bankruptcies, delayed or failed service launches, and antenna siting headaches have all been overcome to allow for the delivery of a service 4 years in the making. Per a Strategis Group report, there were approximately 8.2 million PCS subscribers in the United States as of January 1999. The top 50 wireless markets all had at least one operating PCS carrier, and almost 50 markets had at least two PCS carriers in operation. According to the Strategis Group, there were more than 40 million cellular and PCS subscribers in the United States in 1996, which represents a penetration rate of about 14.5%. Bear in mind that this information is almost 4 years old and that the penetration rate is certainly higher as of December 1998. The Strategis Group also predicts there will be more than 140 million wireless subscribers in the United States by the year 2007. That equates to approximately one wireless phone in place for every two persons in the country!

The existence of the PCS carriers, in and of itself, has had a very beneficial impact on the wireless market as a whole, because the PCS carriers have brought down prices and increased innovation. These are the things that will continue to bring more and more customers into the wireless marketplace.

29.9.2　Cellular ("Incumbent Carriers") versus PCS Carriers: Competition Assessment in 1998

The competitive charge by the PCS carriers has been strong, but there are also some built-in advantages to being an incumbent service provider. Most cellular carriers have bucked industry predictions and successfully migrated their networks to become hybrids of analog and digital systems, with the high-usage customers being moved onto the digital networks.

Cellular carriers haven't sat by idly as the PCS carriers launched their networks. They have significantly ramped up their existing networks with digital technology to make them more PCS-like. The distinctions between PCS and digital cellular service offerings have blurred to the point where the only real difference between the services is the frequency (800 MHz for cellular, 1.9 GHz for PCS). The migration to digital systems has allowed the cellular carriers to not only alleviate congestion on their analog networks by moving their high-end users to the digital systems, but it's allowed them to be more price competitive, in line with PCS service offerings. The migration to digital systems has also allowed the cellular carriers to offer enhanced services such as voice mail, encryption, and paging. Although many industry insiders thought the PCS players would have inherent benefits at 1.9 GHz, analog (cellular) carriers have been able to duplicate those services in the 800-MHz spectrum. "Incumbent" cellular carriers have also found that it is considerably easier to install digital infrastructure on *existing* (cellular) towers than to win approval from local zoning boards to erect new digital towers, as the PCS carriers have to do.

29.9.3   Current Trends in PCS and Cellular

Among the PCS carriers, Sprint PCS has been the most aggressive when it comes to promoting a nationwide "all digital" network. By April 1998, Sprint PCS had captured 37% of the overall PCS subscriber market share, followed by Primeco PCS with 16%, Pacific Bell with 13%, and AT&T Wireless with 7%, according to the Strategis Group. Although PCS coverage may be spotty as carriers still grow their footprints, it is getting better all the time. In October 1998, AT&T took a different expansion route by acquiring Vanguard Cellular. Vanguard provides cellular service under the Cellular One brand name to 625,000 subscribers in rural and suburban markets in the eastern United States. The consolidation trend is evident with other wireless carriers as well. In 1998, Alltel Corporation acquired 360 Degrees Communication (formerly Sprint Cellular). This move allowed for considerable expansion of Alltel's coverage areas.

The issue of community backlash against tower siting (especially for PCS carriers) leads some industry insiders to believe that cellular analog systems won't die the quick death that many people predicted a few years ago. They see a competitive playing field among analog (cellular) and digital (PCS) carriers. Plenty of people have no need to migrate from analog to digital systems. As a matter of fact, 50% of new cellular customers are being signed up on the analog systems. There is no reason to not continue using the analog systems since there is so much money already invested in them. This is especially true in rural America, where many PCS carriers cannot justify a business case to build out these areas due to the types of traffic that exist in rural areas.

Yet digital coverage is improving considerably. It is estimated that PCS services—all-digital systems—are now available to at least 50% of the U.S. population. Some cellular carriers have now discontinued selling analog-only phones and market dual-mode phones only.

29.9.4   Paradigm Shift

Not all carriers are doing this, but in some cases airtime charges are becoming equal to landline rates. Depending on rate plans, the rates can go as low as 8 to 10 cents per minute. Wireless service is starting to become a commodity business. Several carriers have recently developed the marketing concept that they should think in terms of number of customers, not minutes of use or distance traveled on the network (i.e., amount of roaming that customers do). That is why, in May 1998, AT&T introduced its Digital One Rate pricing plan, which charges customers a single, all-inclusive rate anywhere in the United States without roaming or long distance charges. In September 1998, Bell Atlantic introduced a similar pricing program. In October, Sprint jumped on the bandwagon with its own national price plans that range from 600 minutes for $70 per month to 1500 minutes for $150 per month. However, most of these plans are targeted toward high-usage customers.

Carriers are still reluctant to compete using price wars, because their networks are not yet robust enough. The implication is that price wars are practical only if the quality of the product across competitive lines is high enough or similar enough to allow any given competitor to focus intensely on price (or unique features or services). In the case of wireless service, this is not true (has any wireless customer never had a dropped call?). Part of this problem *also* stems from the fact that the networks have capacity problems, because the wireless carriers are starting to reserve capacity for wireless data and more advanced services. As carriers start to focus on applications other than voice, they realize that voice *and* data applications use a considerable amount of bandwidth. Unfortunately, these advanced services are currently unknown and undeveloped, and the network infrastructure is not in place yet to support any nonvoice services that are worthy of strong marketing efforts.

The future of wireless telephony is not clear in terms of technologies and applications, and most users don't care if their service is analog or digital. For those who do prefer digital service, there is little concern over whether the service is CDMA, TDMA, or GSM. Most analysts see a free market for all three of these digital radio technologies as the industry continues to mature. The nagging question today lies in the so-called 3G technologies, where applications using data rates at 2 Mbps or greater are envisioned. A key problem is that the PCS operators are currently operating in the spectrum that other countries have reserved for 3G services, so the PCS carriers would have to obtain additional spectrum in order to offer full-scale services. Until that happens, though, the regulatory bodies, vendors, and carriers alike will have to agree on standards. This process is being addressed by the ITU.

The PCS carriers are attracting more and more of the higher revenue customers because they offer cut-rate pricing. Yet the real challenge to the wireless industry is the issue of billing. With convergence looming over the network plans of all types of public carriers, the question of tracking and billing different types of network traffic becomes paramount.

According to the Telecom Act of 1996, no more than eight wireless carriers can offer service in any given city (two cellular carriers and the six broadband PCS carriers). Due to this constraint and the multiple options for synergies across the lines of different carriers (i.e., continuous coverage potential), analysts anticipate more market consolidations over the next few years, as was demonstrated recently by the AT&T acquisition of Vanguard Cellular, and the Alltel Corp acquisition of 360 Degrees Communications.

Although the future of wireless is uncertain, it is simultaneously brimming with promise. The only other technical platform that has as much or more potential is the development of the new public telephone network, the Internet. The potential, ultimate merging of these two technologies may indeed be the network nirvana of the future.

29.9.5   Convergence

Like the telecommunications industry in general, convergence of voice, data, and ultimately video traffic onto one operations platform will ultimately apply to the infrastructure of wireless carriers as well. This evolution will be likely to occur once the industry has come to the point where coverage and reliability of the wireless network infrastructure is adequate enough to consider convergence.

This migration will occur in the wireless world for the same reasons that it's occurring in the landline telecommunications world: because it makes sense. It offers one platform to manage, one platform to maintain. It decreases training and human resource requirements. It facilitates and streamlines troubleshooting in the network. It allows for less points of failure in the network. It is the direction in which the network providers of the world are marching, and the wireless carriers will join the march at some point in the near future.

29.10   Test Questions

True or False?

1. _____ In March 1998, the company that produces the Swatch line of watches unveiled a wrist phone as a new product.

2. _____ The second-generation wireless service describes all-digital PCS services.

3. _____ One of the eventual goals of most cellular and wireless companies is to completely displace the need for a landline telephone network (by the general public).

Multiple Choice

4. There is a movement which describes efforts by companies and standards bodies to develop a new digital wireless technology that is developed from the major existing digital wireless radio technologies: TDMA, CDMA, and GSM. What is the name that refers to the standard which is the goal of this movement?
   (a) G Force
   (b) NextGen
   (c) 3G
   (d) PCS
   (e) None of the above

5. How does the scanner station (sniffer) operate in wireless PBX systems?
   (a) Scans the PBX for available trunks
   (b) Scans cordless phone frequencies for available channels (frequencies)
   (c) Scans the external macro cellular system to determine which radio frequencies are being used, seeking the least-used frequencies
   (d) Scans the neighboring market's system for unused frequencies

TEST QUESTION ANSWERS

Chapter 1

1. False
2. False
3. False
4. d
5. d
6. d

Chapter 2

1. False
2. True
3. True
4. False
5. False
6. c

Chapter 3

1. False
2. h
3. c
4. c
5. c

Chapter 4

1. True
2. False
3. True
4. True
5. c
6. c

Chapter 5

1. False
2. False
3. False
4. True
5. d
6. g

Chapter 6

1. True
2. False
3. False
4. False
5. g
6. b
7. c
8. c
9. d

Chapter 7

1. True
2. True
3. f
4. c
5. f

Chapter 8

1. True
2. False
3. False
4. d
5. e
6. c

Chapter 9

1. See Figure 9-3
2. False
3. c
4. c
5. a

Chapter 10

1. False
2. True
3. False
4. e

5. b
6. c

Chapter 11

1. False
2. d
3. b
4. c
5. d

Chapter 12

1. True
2. True
3. True
4. a
5. e

Chapter 13

1. True
2. False
3. True

Chapter 14

1. Tracking, call processing, paging, call handoff, billing, roamer data, traffic and call processing stats
2. c

Chapter 15

1. True
2. False
3. f
4. c
5. d

Chapter 16

1. True
2. False
3. h
4. d

Chapter 17

1. True
2. True
3. Section 17.2
4. a
5. d

Chapter 18

1. f
2. d

3. b
4. d
5. b
6. c
7. e
8. c
9. True
10. False
11. False

Chapter 19

1. b
2. c
3. True
4. False
5. True
6. True

Chapter 20

1. a
2. d
3. True
4. False
5. False

Chapter 21

1. c
2. b
3. a
4. d

Chapter 22

1. c
2. c
3. c
4. c
5. c
6. d
7. False
8. False
9. False

Chapter 23

1. c
2. g
3. d
4. c
5. d
6. False
7. False
8. False

Chapter 24

1. True
2. False
3. True
4. b
5. f
6. c
7. c
8. WAE, WSP, WTP, WTLS, WDP
9. See Section 24.10.2

Chapter 25

1. d
2. c

Chapter 26

1. c
2. c
3. c

Chapter 27

1. d
2. c
3. c
4. False
5. True
6. True

Chapter 28

1. True
2. True
3. True
4. False
5. b
6. c

Chapter 29

1. True
2. True
3. True
4. c
5. c

REFERENCES

Allen Telecom Group, Decibel Products Division, "About Base Station Antennas."

Allen Telecom Group, Decibel Products Division, "About Combiners."

Allen Telecom Group, Decibel Products Division, "About Duplexers."

D. M. Balston and R. C. V. Macario (editors), *Cellular Radio Systems*, Artech House, 1993.

Tarre Beach, "Structurally Sound?" *Wireless Review Magazine*, May 15, 2000.

Paul Bedell, "Interconnection Primer," *Wireless Review Magazine*, June 15, 2000.

Jose Bernal, "Smart Antenna Systems," DePaul University TDC 512 Case Study, June 2000.

Jerry Blake, "MSS Carriers Beam Service to a Variety of Customers," *Radio Communications Report (RCR)*, January 29, 1996.

Pat Blake, "The Space Race," *Telephony Magazine*, November 20, 1995.

Karissa Boney, "Appeasing the Locals," *Wireless Review Magazine*, January 15, 1998.

Orie Carter, "Wearable Cellular Phone," *Chicago Tribune*, January 4, 1996.

Cellular Business Magazine, "Subscription Fraud: The Original Wireless Fraud," September 1995.

Cellular Telecommunications Industry Association, *The Wireless Fact Book*, Spring 1995.

Cellular Telecommunications Industry Association web site, "The History of Wireless" (www.wow-com.com).

Teri Lynn Chesek, "GSM," DePaul University TDC 512 Case Study, March 2000.

Chicago Tribune, "Bells Sounding over Cell Phone Theft, Fraud," December 25, 1995.

Chicago Tribune, "High Tech Bet—Cellular's Success Makes New Technology Seem a Surer Thing," December 5, 1994.

Chicago Tribune, "Local Phone Wars Just One Call Away," April 4, 1995.

Narisa Chu, "LMDS Technology, Deployment and Application," March 1998 (www.cwlab.com/whitepapers/lmds.htm).

John S. Csapo, "The Power of Being Small," *Telecommunications Magazine*, September 1999.

Larry Dvorkin, "CDMA Technology," DePaul University TDC 512 Case Study, June 2000.

Stephen Dynes, "Planning Microwave Radio Networks," Digital Microwave Corporation (DMC), July 1995.

Philamkelly Empleo, "Case Study in Emergency 911," DePaul University TDC 512 Case Study, June 2000.

Tom Flynn, "CDMA Technology: An Overview," DePaul University TDC 512 Case Study, June 2000.

Nancy Gohring, "CPP Strikes Again," *Telephony Magazine*, April 3, 2000.

Nancy Gohring, "Location, Location, Location," *Telephony Magazine*, May 11, 1998.

Betsy Harter, "Ready for Inspection," *Wireless Review Magazine*, May 1, 1999.

Richard Henderson, "The Key to Capacity," *Telephony Magazine*, June 1, 1998.

Donna Hildebrand and Char Chase, "Wireless PBX: An Overview," United States Cellular Corporation Advanced Wireless Group presentation, January 1996.

Hybrid Networks, "An Introduction to Fixed Broadband Wireless Technology" (www.hybrid.com).

International Engineering Consortium, "LMDS Tutorial" (www.iec.org).

International Engineering Consortium Web ProForum, "Wireless Intelligent Network (WIN) Tutorial," June 2000.

Frank James, "Information Age Gets New Rules of Road," *Chicago Tribune*, February 2, 1996.

Tadeusz Klos, "On the Microwave Path," Digital Microwave Corporation (DMC), December 1996.

Rikki Lee, "Carriers Feel No Fear from PCS," *Wireless Week Magazine*, January 15, 1996.

George Leopold and Brian Santo, "Local Multipoint Distribution Service: Wireless Wonder or Broadband Bust?" February 11, 1998 (www.techweb.com).

Asha Mehrotra, *Cellular Radio: Analog and Digital Systems*, Artech House, 1994.

Jason Meyers, "Calling All Carriers," *Telephony Magazine*, September 15, 1997.

Jason Meyers, "Mission Accomplished," *Telephony Magazine*, August 2, 1999.

Jason Meyers, "Stumping for Digital," *Telephony Magazine*, May 13, 1996.

Roy Moore, "Do Your Homework—Selecting Towers," *Communications Magazine* (FWT Inc.), February 1995.

Paul Nevill, "The SS7 Migration," *Wireless World Magazine*, April 1997.

Susan O'Keefe, "Upward Mobility," *Telecommunications Magazine*, December 1997.

Kelly Pate, "Carriers Combating 'Clone 'N Roam' Fraud in Markets," *Radio Communications Report (RCR)*, February 26, 1996.

Radio Communications Report (RCR), "Broadband PCS Auction Winners and the Markets They Won," March 27, 1995.

Niraj Shah, "Microwave Path Design," *Wireless Review Magazine*, September 1, 1998.

Sprint Broadband Group, "MMDS—Better Than Sliced Bread," (www.sprintbroadband.com).

Raymond Steele, "What Is CDMA?" *Mobile Communications International Magazine*, Multiple Access Communications Ltd.

Thomas Stroup, "The Hurdles of Microwave Relocation," *Cellular Business Magazine*, November 1995.

Telephony Magazine, "Tower Space Proliferates," September 20, 1999.

Karissa Todd, "The WLL to Succeed," *Wireless Review*, May 15, 1998.

Jon Van, "Chicago Market Brings Highest Wireless Bids," *Chicago Tribune*, March 14, 1995.

Jon Van, "The Drive to Pinpoint Mobile Callers," *Chicago Tribune*, June 26, 1998.

Jon Van, "Local Phone Calls Just One Call Away," *Chicago Tribune*, April 4, 1995.

WCA International, "MMDS Overview" (www.wcai.com).

Wireless Data Forum web site (www.wirelessdata.org).

Wireless Review Magazine, "Shelter So Fair," April 1, 2000.

Wireless Review Magazine, "Stats Jeopardy," December 15, 1999.

"WLL Systems," PHS MOU Group Information web site, January 1998.

Ronald E. Yates, "Popularity of Cell Phones Far Exceeds Expectations," *Chicago Tribune,* April 1994.

Deborah Young, "Ready to Roam?" *Wireless Review Magazine,* May 15, 2000.

John Yuzdepski, "A WAP Primer," *Wireless Review Magazine,* April 1, 2000.

WIRELESS INDUSTRY AND TELECOM PUBLICATIONS AND INTERNET SITES

Publications

RCR (Radio Communications Report) magazine, Denver, Colo. Published weekly. For subscription information, call 800-678-9595.

Wireless Review, published twice monthly by Primedia Intertec Publications, 9800 Metcalf, Overland Park, KS 66212-2215.

Mobile Phone News, Philips Business Information, Potomac, Md. Published weekly. For subscription information, call 301-424-4297. (Philips Business Information also publishes the following: *PCS News, Land Mobile Radio News, Wireless Data News, Mobile Satellite News, Wireless Business and Finance, Wireless Industry Directory, Mobile Product News,* and *Global Positioning and Navigation News.*)

Cellular Business Magazine, published monthly by Intertec Publishing, 9800 Metcalf, Overland Park, KS 66212-2215.

Wireless World Magazine, published quarterly by Intertec Publishing, 9800 Metcalf, Overland Park, KS 66212-2215.

Mobile Office Magazine, Woodland Hills, Calif. Published monthly. For subscription information, call 818-593-6100.

Telecommunications Magazine, www.telecommagazine.com.

Telephony Magazine, Intertec Publishing, 55 E. Jackson, Chicago, IL. Published weekly. For subscription information, call 800-441-0294.

Wireless Magazine. Call 800-915-0999 or e-mail thutten@wirelessmag.com.

Wireless Week Magazine, www.wireless.com.

America's Network Magazine. Call 714-513-8400 or check web site www.americasnetwork.com.

Internet Sites

www.wow-com.com   Web site of the Cellular Telecommunications Industry Association (CTIA).

www.pcia.com   Web site of the Personal Communications Industry Association

www.nacn.com   Web site of the North American Cellular Network

www.nextel.com   Web site of Nextel Corporation, E/SMR service provider

www.wirelessdata.org   Web site of the Wireless Data Forum

www.illuminet.com   Web site of Illuminet, nationwide SS7 network provider

www.fcc.gov/wtb   Web site of the FCC's Wireless Telecommunications Bureau

www.teledesic.com   Web site of the Teledesic satellite PCS system

www.globalstar.com   Web site of the Globalstar satellite PCS system

www.hybrid.com   Contains white papers on wireless topics

www.webopedia.lycos.com   Web site with over 4000 computer and technical terms defined

www.whatis.com   On-line technical dictionary/encyclopedia

www.iec.org   Web site of the International Engineering Consortium. WebProForum On-Line Tutorials are housed on this site, on a huge array of network-related topics and technologies. New tutorials are added on a regular basis.

www.techguide.com   Web site with large listing of white papers on network-related topics

INDEX

About the Author

Paul Bedell obtained an M.S. degree with distinction from the Telecommunications Management Program at DePaul University in Chicago, one of the leading programs of its kind in the United States. In 1995, he designed and has since taught the Cellular and Wireless Telecommunications course at DePaul, in the School of Computer Science, Telecom and Information Systems.

Bedell began his telecommunications career in the U.S. Army Signal Corps, serving in West Germany from 1985 to 1988, where he worked as a multichannel communications equipment operator at a remote signal site. After receiving his discharge in 1988, Bedell spent 5 years working for several *Fortune* 500 companies as a telecommunications analyst. From there, he moved to the wireless industry, where he spent 5 years working for three leading wireless carriers, both cellular and PCS.

Bedell spent $2^1/_2$ years as a network engineer for United States Cellular. During that time, he designed and implemented fixed and interconnection networks in most regions of the United States. He then moved to PrimeCo Personal Communications, where he built out the prelaunch fixed network for the metro portion of the Chicago MTA. From there he took a position with Aerial Communications (since purchased by VoiceStream Wireless), where he designed and built out Aerial's prelaunch wide-area network (WAN), which connected the company's eight MTAs to its network operation center (NOC) in Tampa, Florida. The WAN supported network management for GSM network elements, IT network elements, and 611 traffic to Customer Care, also based in Tampa. Bedell then moved to Ameritech's long-distance business unit, Ameritech Communications, Inc. (ACI), where he managed the implementation of ACI's 42-node data network that supported its 2500-mile SONET ring that traversed the midwestern United States. He also installed two Voice over IP networks at ACI before moving into product management after the merger with SBC was completed. He is currently Associate Director of Product Management in the Optical Data Network Group at SBC (Ameritech) in Chicago.

He can be reached at pbedell@megsinet.net.